CONTENTS

Illustration on Cover:
    Summer Holly *(Comarostaphylis diversifolia)*

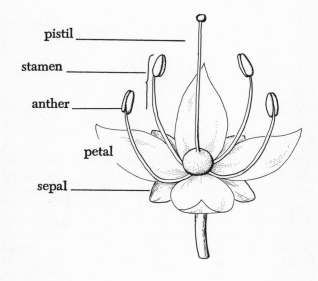

pistil _____

stamen _____

anther _____

petal

sepal _____

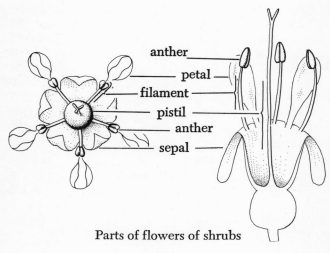

anther _____

petal _____

filament _____

pistil _____

anther _____

sepal _____

Parts of flowers of shrubs

[4]

CHARLES W. LOOMIS

California Natural History Guides: 15

NATIVE SHRUBS

OF

SOUTHERN CALIFORNIA

BY

PETER H. RAVEN

UNIVERSITY OF CALIFORNIA PRESS

BERKELEY AND LOS ANGELES 1970

UNIVERSITY OF CALIFORNIA PRESS
BERKELEY AND LOS ANGELES, CALIFORNIA
UNIVERSITY OF CALIFORNIA PRESS, LTD.
LONDON, ENGLAND
© 1966 BY THE REGENTS OF THE UNIVERSITY OF CALIFORNIA
SECOND PRINTING, 1970
STANDARD BOOK NUMBER 520-01050-7
LIBRARY OF CONGRESS CATALOG CARD NUMBER: 65-28527
PRINTED IN THE UNITED STATES OF AMERICA

INTRODUCTION

Southern California, with its valleys, high mountains, and deserts, is exceptionally rich in native shrubs. Much of the area is too arid for the growth of trees, and extensive brushland communities often take the place of the forests of more temperate regions. The chaparral, a dense tangle of hard-leaved, evergreen, often spiny shrubs, covers thousands of acres on our mountains, and the periodic fires that range through it are a source of consternation to the area's human inhabitants. Fires have periodically burned through the shrub communities of southern California for millions of years, and, as we shall see, elicit various responses from different native shrubs.

Many of our shrubs are important in the gardens of the world, and others, hitherto little-known or neglected, deserve much wider planting than they have been accorded. Many of the problems of gardening in our region are the result of attempting to recreate in arid southern California gardens the environments of, for example, northern Europe, or Hawaii, together with a full complement of the plants of those very different regions. Our own native plants, which have lived in the region for thousands of years, are well adjusted to its climate and need much less care than many of the exotics. In addition many of them are outstandingly beautiful and deserving of wide familiarity and planting in gardens. A very useful book for those interested in pursuing this subject further is *Native Plants for California Gardens,* by Lee W. Lenz, published by Rancho Santa Ana Botanic Garden, Claremont.

For the purposes of this book, southern California is considered to include Santa Barbara, Ventura, and Los Angeles counties, and San Bernardino County south of U. S. Highway 66, as well as the more southern Orange, Riverside, San Diego, and Imperial counties. Within this richly diversified area grow approximately 400 kinds of shrubs, the great majority of them mentioned in this book. At the end of the book is a check list of the shrubs that are mentioned here, and book references that will help in the identification of shrubs not covered here.

Black-and-white illustrations are the work of Jean Colton, except for the figure on p. 4 and the figure on p. 34, which were drawn by Kay Brown. The color illustrations are by Eugene Murman. Both the author and the publisher wish to express their thanks to Dr. Lawrence Cark Powell and the University of California Library, Los Angeles, for permission to reproduce in reduced form the Murman watercolor drawings, which are part of the special collections in that library.

WHAT IS A SHRUB?

The answer to this question can be determined only by deciding what is a tree and what is an herb. These are somewhat arbitrary classes, but for our purposes a shrub will be defined as a plant with a distinctly woody trunk arising from the ground (not sprawling on it). Woody plants that usually reach a height of 12 to 15 feet are considered trees, and are not included here, although some trees that rather frequently have shrubby forms are mentioned incidentally. Southern California trees are treated in *Native Trees of Southern California*, by P. Victor Peterson, University of California Press. Cacti are not treated in this book.

THE NAMES OF SHRUBS

Like all other living organisms, shrubs are given names which consist of two parts, the genus and the species. These scientific names are in Latin and are standard for a given organism throughout the world. For example, Chamise, a common shrub of our brushy slopes, is called *Adenostoma fasciculatum*, *Adenostoma* being the genus, *fasciculatum* the species, and the two together, the scientific name of the plant. A related species, Red Shanks, is *Adenostoma sparsifolium*. These two species are grouped together in the genus *Adenostoma* because they have a great many characteristics in common, but, because they are different kinds, are considered two different species. In addition to the Latin name, many shrubs of southern California have an English name, usually given by the inhabitants of a particular place or by the author of a book. Unlike the scientific names, these English names vary from locality to locality and are not widely understood. Because many of our southern California shrubs lack standard English names, we have emphasized the scientific names in this book. In view of the many scientific names widely used and understood—such as Chrysanthemum, Sequoia, and Ceanothus—this should cause little problem. Whenever English names are available, they are mentioned also.

Genera are grouped into larger units called families, but these are not stressed here, except in the check list at the end. For example, *Adenostoma* is grouped with such genera as *Prunus,* the plums, and *Fragaria,* strawberry, into the Rose Family. The families are then grouped into larger and larger units, the members of which have some characteristics in common, but not as many as, say, the members of a particular genus. Our shrubs are subdivided into two great groups, the Gymnosperms, consisting of the conifers and related groups of plants, and the Angiosperms, consisting of flowering plants. Only two of our shrubby genera are Gymnosperms: *Ephedra,* or Mormon Tea, and *Juniperus,* juniper. The Angiosperms are likewise further subdivided into two major groups, Monocotyledons, including grasses, lilies, palms, and, among our shrubs,

[7]

Yucca and *Nolina;* and Dicotyledons, including the rest of our shrubby genera. The distinctions between these major groups are fundamental, but they will not be dwelt on here, since they are not of practical importance in identifying our shrubs.

NATURALIZED SHRUBS

In addition to the hundreds of native shrubs in southern California, a few have been brought in by man and, seeding themselves, have become established as members of our flora. For example, Castor Bean, *Ricinus communis,* and Tree Tobacco, *Nicotiana glauca,* were both introduced into California as garden plants in Spanish days and have now become established as members of our flora. Tamarisks were brought in for planting far more recently. The few shrubs from elsewhere that have become naturalized citizens of our roadsides are treated in full here, but the many hundreds of others that are planted in our parks and gardens, surviving only with the aid of man, are not mentioned. A comprehensive guide to the woody ornamental plants of southern California is badly needed, but an adequate one is unfortunately not available. This is due in large measure to the rapid changes in the roster of our cultivated plants as tastes change and new plants become available in the horticultural trade.

IDENTIFYING SHRUBS

A limited knowledge of terminology is necessary for the identification of our native shrubs. In this book the terminology has been simplified as much as is consistent with accuracy. A basis for the terms used is given in the following discussion, which should be supplemented by reference to the illustration on page 4 and the pictures of individual shrubs scattered throughout. In all these pictures, the small line equals one inch.

Aside from two groups of cone-bearing plants—*Ephedra* and *Juniperus*—our native shrubs bear more or less conspicuous flowers. Since their classification is based largely on the structure of the flower, it is important to understand something about it. A flower has in its center a *pistil* or group of pistils usually consisting of a swollen lower portion and a stalk, at the top of which the pollen is deposited. When the outer parts of the flower have fallen off, following fertilization, the pistil matures into the *fruit*. It is very important to remember that when a botanist uses the term fruit he has in mind not only the familiar things we would commonly call such, but all seed-containing structures that mature from the pistil. Thus, botanically, a bean pod, an acorn with its cup, and a berry are all fruits. The pistil or group of pistils is surrounded by a circle of few to many *stamens,* each of which has at its summit a pollen-bearing structure called the *anther.* The dust-like pollen is carried from the anthers to the pistil. There, a fine tube grows from it into the pistil, where fertilization takes place. This results in the production of seeds which are the start of the next generation. In our shrubs with inconspicuous greenish flowers, such as oaks and willows, the pollen is carried about by the wind. In others, like Ceanothus or Fremontia, with bright-colored flowers, pollen is usually transported by various insects. These insects visit the flowers to feed on the pollen or the honey-like nectar (which in many plants is secreted by glands at the base of the flower). Some of our shrubs with bright red flowers are visited by one or more kinds of hummingbirds. These bright aerial acrobats may also occasionally go to almost any reasonably conspicuous flower. While probing for nectar or for tiny insects in the flowers, these birds often pick up pollen on their throats, faces, or bills, and then carry it to another flower.

The majority of our shrubs, and those with relatively showy flowers, have both stamens and pistils in each flower. In some of our shrubs, the flowers are either *staminate,* therefore having no pistil, or *pistillate,* having no stamens. In willows, staminate and pistillate flowers are never found on the same bush. A given plant has flowers of only one sort. In Burrobrush, *Franseria dumosa,* on the other hand, both staminate and pistillate flowers occur on the same bush, but in different places.

Outside the circle of stamens are often two more whorls of flower parts, the brightly colored *corolla* and the often greenish *calyx.* If they are separate, the individual members of the corolla are called *petals,* and the individual members of the calyx, *sepals.* When we look at an unopened flower, we normally see the sepals, joined together in the bud. Upon opening, this outer whorl splits open to reveal the petals and inner flower parts. In some of our shrubs, as in Toyon, *Heteromeles,* the petals are separate, whereas in others, as in sages, *Salvia,* they are joined together into a tube. Other shrubs, like willows, lack petals entirely, and have the stamens immediately inside the sepals. Although the corolla usually is the conspicuous part of the flower, this is not necessarily so. In Ceanothus, for example (see page 4), it is the bright blue calyx and anthers that add most of the color to the showy flower clusters, while the narrow hooded petals are relatively inconspicuous. A number of plants have a tube-like calyx, with the petals and stamens attached at its summit; an example of this kind of flower, *Ribes,* is shown on page 4.

In describing the vegetative parts of shrubs in this book, the following terms are used. On a stem, the point where a leaf is attached is called a *node.* Depending on the kind of shrub, there may be one, two, or several leaves at a given node. If several leaves arise from the same point, the leaves are said to be *whorled.* A more common condition is one in which two leaves, forming a pair, arise on opposite sides of the stem from a single point; they are said to be *opposite.* Finally, and most common among our shrubs, is the condition in which the leaves are said to be *alternate;* a single leaf arises at each node. Taken together, the leaves are arranged in a spiral fashion up and down the stem.

Some shrubs have leaves that are divided all the way to the midrib into *leaflets.* Such leaves are said to be *compound.* Examples are ash, lupine, rose, and walnut. Leaves that are deeply divided, but not cut all the way to the midrib, are said to be *lobed,* and those that have a shallowly cut margin are *toothed.* If the margin is not cut, the leaves are *entire.*

Like many trees, some shrubs lose all their leaves in the winter, a complete new crop growing in the spring. They are called *deciduous.* Other shrubs lose their leaves sporadically throughout the years, and the plants never are completely bare. Thus they are *evergreen* shrubs. Unfortunately, the term evergreen is sometimes incorrectly restricted to the conifers.

ACTIVITIES

There are so many kinds of native shrubs in southern California that learning to recognize them is a rewarding nature activity in itself. In gaining this familiarity, many interesting sidelines can be explored.

VISITING BOTANIC GARDENS

The citizens of southern California are fortunate in having two excellent botanic gardens that are restricted to native plants. At Rancho Santa Ana Botanic Garden, 1500 North College Avenue, Claremont, or in the Santa Barbara Garden, 1212 Mission Canyon Road, Santa Barbara, nearly all the shrubs in this book can be observed. Not only are these living collections invaluable for learning the shrubs, but they demonstrate graphically how these shrubs can be used in the home garden. These gardens will be glad to furnish information on where native shrubs can be purchased from nurseries. Native plants can be dug up in the wild on private property with the written permission of the owner. Usually they can be transplanted only with difficulty! It is far better to purchase healthy nursery stock for your garden.

BECOME FAMILIAR WITH SHRUBS

There is far more to learn about our native shrubs than simply their names. Even in a small area, the different kinds of shrubs usually grow in very different surroundings. One kind may be on steep, south-facing rocky slopes, whereas on the level flats at the base of these slopes it may be entirely replaced by a second species. It is fascinating to work out in detail the relationships between the shrubs and their habitat. In several of our most common groups of shrubs, for example, manzanitas, Ceanothus, sages, and oaks, hybrid individuals may be found where two species are growing together. The production of such hybrids is discussed further under the groups where they are known. It is particularly interesting to search for such situations and then see if there are hybrids. Hybrids are easy to recognize by their intermediate characteristics. Usually they occur in small numbers in relation to the numbers of individuals of the two species. When they are between two species as different as White Sage, *Salvia apiana,* and Black Sage, *S. mellifera,* they are very conspicuous.

When hybrids are found, we can study the requirements of the two parent species and observe what sorts of places the hybrids are growing in. If two species occur in somewhat different situations, a few hybrids may often be found where the preferred conditions of the parents coincide—say, a north-facing slope and an east-facing one, with hybrids on the northeast side.

If, in a given area, we find forest, chaparral, and grassland, for example, growing together, what are the reasons for one plant to be in a particular place and not the other? The slope, its direction, the local rainfall, the kind of soil—all these and many other factors may be of importance in determining where plants grow. Even the scientists who study such matters still have a great deal to learn about them.

COLLECTING SPECIMENS

Small branches of shrubs in flower or fruit can be dried between the pages of a weighted-down book and then glued onto cards or the pages of a notebook. Individual leaves can also be pressed in this way, and these specimens form an interesting nature-study record. The names of the shrubs and of the places where the specimens were collected can be written on the pages. But remember that specimens can be taken only on private land and with the permission of the owner, and never in national and state parks or monuments; local regulations may also apply.

AFTER A FIRE

The fires which periodically ravage our brushy hillsides are a continual source of danger and economic loss. It can be extremely interesting, however, to study the reaction of native plants to them. Some grow freely from seed following the burn, whereas others send up tender new shoots from underground parts. A careful observer can visit a newly burned slope periodically and learn how the various kinds of native plants have recovered from the fire. Often the new seedlings have leaves entirely different in form from the adult leaves of the same plant. For example, seedlings of Chamise, *Adenostoma fasciculatum*, have soft, deeply lobed leaves instead of the hard needle-like ones found on the adult plant. This can add interest to the study. But as you travel about to study the native shrubs, be careful of fire yourself, particularly during the long dry season!

The relation between our native plants and fires is manifested by so many special modifications in the plants that it is clear that fires have always been a feature of these brushy communities. Similar types of vegetation have developed under the influence of fire in South Africa, in Chile, in Australia, and else-where—all regions at about the same latitude as southern California, and with similar summer-dry climate.

NATIVE SHRUBS OF SOUTHERN CALIFORNIA

In this book, the native shrubs are treated in eleven different groups. To decide to which group an unknown shrub belongs, read through the following list in order until you find a description which applies *fully* and *exactly* to your unknown. Then turn to the page indicated and work on from there.

GROUP I

Branches jointed, with 2 or 3 scale-like leaves at each node, the nodes widely separated; plants bearing cones.

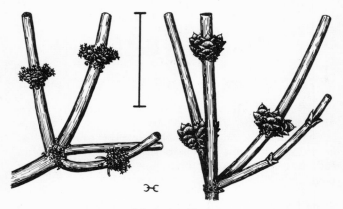

Mormon Tea

Mormon Tea *(Ephedra)*

A group of four closely similar species of twiggy shrubs with stout greenish or grayish-green branches coming from shaggy, thick central trunks. Three of the species are confined to the desert; the fourth, *Ephedra californica,* extends to the coast near San Diego. Seed-bearing cones, which are scaly and contain two nut-like seeds, are borne on some of the plants, pollen-bearing cones on others. Both kinds can usually be found growing together in a colony. The branches are often swollen by insect galls. Two of the species have their leaves in groups of 2 at a node. *Ephedra viridis* has bright green slender erect branches. It occurs on mountain slopes and in washes on the Mojave Desert. *Ephedra nevadensis* has dull, dark green branches which are stout and spreading. It usually occurs at lower elevations than *E. viridis* and ranges from the Mojave Desert southward to the western margins of the Colorado Desert in San Diego County.

The other two kinds of Mormon Tea in southern California have 3 leaves at each node. One of them,

E. californica, is our most common species, occurring throughout the deserts, west to the coast near San Diego, and northward in the dry inner Coast Ranges. *Ephedra trifurca,* which is like *E. californica* in having 3 leaves at each node, differs from all other southern California species in having the branches terminating in stout spines. It is very rare, being known only from a few widely scattered localities on the Colorado Desert and from near Daggett.

GROUP II

Branches densely covered with overlapping appressed scale-like leaves, less than 1/16 inch long.

As we have constituted it, this group is a very artificial one, since it brings together the native California Juniper, which is a conifer, with the introduced Tamarisk, a Dicotyledon.

California Juniper *(Juniperus californica)*

The resinous, fragrant large shrubs and small trees of California Juniper cover thousands of acres on the dry inner slopes of most of the principal ranges and along the margins of the desert. The fleshy round cones are about ½ inch thick, and when young are covered with bloom like a plum. Other kinds of junipers occur within the borders of southern California, but are almost always distinctly tree-like.

Tamarisk *(Tamarix)*

These shrubby or tree-like natives of the Old World, especially frequent in the arid regions of the Near East, have been widely planted in our own deserts and dry hills, and have flourished. In many places they have become naturalized, reproducing from seed freely and without the aid of man. Two species are frequently recorded as established in such situations in · southern California, and they may be expected almost anywhere throughout the area. In

[17]

Tamarix pentandra the flower parts are in fives: 5 petals, 5 stamens, and so forth. Its tiny pink or whitish flowers are densely bunched in lateral clusters 1 to 2 inches long, and they may be found on the plant most of the year, although they are most common in summer. *Tamarix tetrandra* has similar flowers, similarly bunched, but they appear on the branches in early spring before the leaves begin to grow. Its flower parts are in fours. Both species of tamarisk are shrubs or small trees, mostly 3 to 15 feet tall.

GROUP III

Leaves long, sharp-pointed, and swordlike or grass-like, the veins parallel to the edges and running the length of the leaf.

Our only shrubby representatives of one of the two great divisions of flowering plants, Monocotyledons, are found in this group. Most of the shrubs of southern California are Dicotyledons. The only exceptions are the two genera grouped here, *Yucca* and *Nolina*. Botanically, they are both considered members of the Lily Family, and of course they are among the most conspicuous features of our flora. The nolinas, which have narrow grass-like leaves and flowers less than ¼ inch long, will be treated first, and then the yuccas, in which the flowers range from ¾ inch to 3 inches in length. All four species of *Nolina* found within our limits have distinct woody trunks and hence are included here, but none is common.

Bigelow Nolina *(Nolina bigelovii)*

In this species a dense cluster of leaves, each about 3 feet long, is borne atop a branched woody trunk that ranges up to about 3 feet in length. The leaves are about ½ inch to slightly more than 1 inch long. The flower stalks, which bear a dense long cluster of tiny white flowers, are up to more than 8 feet long, and, like the other yuccas and nolinas, appear in late spring and early summer. Bigelow Nolina occurs sporadically along the western and northern margins of the Colorado Desert and in a few mountain ranges of the eastern Mojave Desert. Its leaves have shredding brown fibers along their margins, whereas in the closely similar *Nolina wolfii,* Wolf Nolina, the leaves are minutely toothed along the margins and do not have shredding fibers. Wolf Nolina occurs locally from 3,500 to 6,000 feet elevation in the Kingston, Eagle, and Little San Bernardino mountains, and also in the southeastern San Jacinto Mountains.

Parry Nolina *(Nolina parryi)*

Parry Nolina can be distinguished by its narrower, toothed, nonshredding leaves, which scarcely exceed ½ inch in width; otherwise it is quite similar to Bigelow Nolina. Unlike that desert species, however, it occurs in chaparral and along the upper margins of the pinyon pine—juniper woodland. It occurs in Ventura, Riverside, Orange, and San Diego counties, and along the desert slopes of the Santa Rosa and San Jacinto mountains. The fourth kind of nolina in southern California, *Nolina interrata,* is very rare, being known only from chaparral-covered slopes west of the Dehesa School, 8 miles east of El Cajon, San Diego County, where it is locally common. It has narrow leaves like Parry Nolina, but has essentially no woody stem beneath the leaf cluster, this stem being buried under the ground and the clustered leaves borne at ground level.

Mojave Yucca, Spanish Dagger *(Yucca schidigera)*

This is the only southern California yucca that can be considered a true shrub, having distinct woody stems below its clustered, dagger-like leaves. These leaves are yellow-green, about 1 to 3 feet long, and have conspicuously shredding pale marginal fibers. They are trough-like with the margins folded up. The creamy white flowers are about 1 inch or more in length and borne in clusters 1 to 2 feet long. Mojave Yucca occurs over the entire Mojave Desert and south through the San Jacinto and Santa Rosa Mountains and west to the coast in the vicinity of San Diego. It is the commonest yucca in our deserts. The strong fibers of the leaves were used by the Indians for weaving various articles. The other kinds of yucca in our area either lack trunks or, like the Joshua Tree, *Yucca brevifolia,* are distinctly tree-like.

Moths of the genus *Pronuba,* which occur throughout the entire range of *Yucca,* carry pollen to the stigmas of the flowers and then lay eggs in the flower they have pollinated. Their larvae live in the developing fruits, but eat only a percentage of the seeds, leaving the others to germinate and grow into new yucca plants.

GROUP IV

Vines.

Here are included all the twining woody plants of southern California. There are five genera in this group, two of which, *Lonicera* and *Vitis,* have simple leaves; the other three, compound (divided into separate leaflets). In *Lonicera* the margins of the leaf are entire; the vine-like species are treated with the erect ones on pages 62-63. *Vitis,* Grape, has the familiar cut and lobed leaves of the cultivated grape. The remaining three genera have compound leaves divided to the midrib into separate leaflets. They are *Clematis, Rhus,* and *Rubus.*

Pipe-stem Clematis *(Clematis lasiantha)*
The three southern California species of *Clematis* are clambering woody vines which attach themselves by twining thread-like tendrils. They lack petals, but the conspicuous white sepals give the impression of petals. In *C. lasiantha* the sepals are about 1 inch or a little

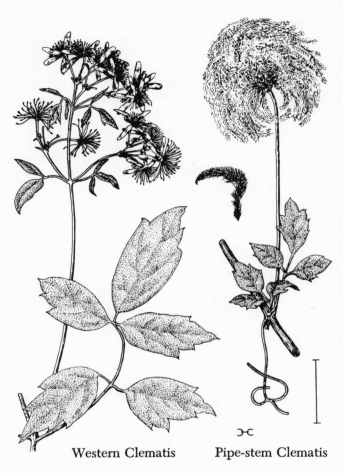

Western Clematis   Pipe-stem Clematis

less in length. The most conspicuous feature of the
vines of *Clematis,* however, are the fruits. Each flower
gives rise to a cluster of small hard fruits each of
which has a long hairy white tail 1 inch or more long.
Thus, in late summer and fall, when the vines are in
fruit, these fuzzy white clusters attract a great deal of

attention. In England a similar species has been given the name "Traveller's Joy." In *C. lasiantha* each leaf is divided into 3 coarsely toothed leaflets. The flowers are in bunches of 1 to 3 on each stalk, and the seeds are covered with short hairs. Pipe-stem Clematis occurs throughout the coastward mountains of southern California.

Southern California Clematis *(Clematis pauciflora)*
Similar to the preceding in having only 1 to 3 flowers on each stalk, Southern California Clematis differs in that each leaf is divided into 5, 7, or 9 leaflets (very rarely). Its seeds are smooth, not hairy. The sepals are about ¼ to ½ inch long. It occurs from the coastward ranges of San Diego County (where common) northward to the mountains surrounding the Los Angeles Basin (where infrequent).

Western Clematis *(Clematis ligusticifolia)*
This long scrambling vine is very different from the other two southern California species of *Clematis* in several respects. It blooms in spring like the others, but its flowers are clustered and its sepals are less than ½ inch long. Its leaves are divided into 5 or 7 leaflets and its seeds are hairy. Whereas the other two species tend to grow on dry brushy slopes, Western Clematis is often found near streams or growing up trees in fertile bottoms. It occurs throughout the mountains of southern California, east to the margins of the desert.

Poison Oak *(Rhus diversiloba)*
This common shrub, which occasionally becomes vine-like and twines up tall trees, may be recognized by its 3-parted leaves, small greenish flowers, and smooth, berry-like white or brown fruits about ¼ inch in diameter. On contact with the skin, it produces in many people a violent allergic reaction. (See also p. 30.)

Blackberry, Raspberry *(Rubus)*

The vine-like members of this genus may be distinguished from other southern California vines by their prickly steams. They are treated on pages 31-33.

Wild Grape *(Vitis girdiana)*

Wild Grape is a long-stemmed woody vine that scrambles over trees and other vegetation near streams and in canyon bottoms. The leaves and black berries are the same as those of the cultivated grape. The flowers of our native grape are minute, greenish, and clustered. The clusters arise on the side of the stem opposite the leaves, as do the twisted and branched tendrils. Wild Grape occurs on the margins of the desert from San Diego County north, and in the Coast Ranges north to Santa Barbara County.

GROUP V

Parasitic shrubs, growing on branches of trees or other shrubs.

Included in this group are mistletoes belonging to the genus *Phoradendron*. The name of the genus comes from two Greek words signifying "thief of the tree." Another genus found within our limits is *Arceuthobium*, Pine Mistletoe, but its species are not woody. They are not completely dependent on the host plant for nourishment, as they have chlorophyll and produce part of their own food. Branches of our species, with their white berries, are often sold at Christmas time. The seeds are covered with a sticky substance by which they adhere to the branches of their host. All five species found in California occur within our limits. They are easily distinguished by the characteristics given below. All have inconspicuous greenish flowers that lack petals and are sunken in the stem and closely bunched.

[23]

Cypress Mistletoe *(Phoradendron bolleanum)*

A mistletoe with well-developed narrow leaves that occurs on firs, junipers, and cypresses throughout the higher mountains of southern California.

Juniper Mistletoe *(Phoradendron juniperinum)*

Juniper Mistletoe, like Cypress Mistletoe, grows on conifers (junipers, incense cedar), but has highly reduced scale-like leaves. It occurs in the San Gabriel, San Bernardino, and San Jacinto mountains, and in the mountains of San Diego County.

Mesquite Mistletoe *(Phoradendron californicum)*

Mesquite Mistletoe has scale-like leaves and occurs throughout the deserts of southern California, growing on various trees and shrubs in the Pea Family. This species has small coral-red berries. Both of the other species of *Phoradendron* that occur on broad-leaved, nonconiferous trees and shrubs have well-developed leaves. They are the following:

Bigleaf Mistletoe *(Phoradendron tomentosum)*

Bigleaf Mistletoe grows on cottonwood, sycamore, willow, and ash, from the Santa Monica Mountains south to San Diego County. It flowers from approximately December through March.

Oak Mistletoe *(Phoradendron villosum)*

Oak Mistletoe grows primarily on various oaks, but also on other broad-leaved trees. Its mature leaves are hairy and rarely more than 1½ inches long, whereas those of Bigleaf Mistletoe are quite smooth and often as much as 3 inches long. Oak Mistletoe occurs throughout the coastward mountains of southern California, and flowers primarily from July through September. Thus its white berries are common in the Christmas trade.

Shrubs with compound leaves, i.e., subdivided to the midrib into leaflets.

This rather large and complicated group contains most of the southern California representatives of the Pea Family as well as a number of other common shrubs. Members of the Pea Family may be recognized by their fruits, technically known as legumes, which are flattened and split along both edges. In addition, many members of this family have highly irregular flowers like those of the sweet pea or garden bean. All plants of the Pea Family, even those with simple leaves, are treated on page 33.

In Group VI, the following hints may be given for identification. In *Berberis* the leaflets have spiny margins, and the flowers are yellow. *Rosa* and *Rubus* have prickles or thorns on the stems. *Larrea,* Creosote Bush, has each leaf divided into only 2 leaflets and has medium-sized bright yellow flowers. *Rhus* and *Juglans* (Walnut) have inconspicuous greenish or greenish-white flowers; *Fraxinus, Sambucus,* and *Chamaebatia* have conspicuous flowers with white petals. Finally, *Isomeris,* a strong-smelling shrub, has yellow flowers and inflated pods connected to the calyx by a stalk. These shrubs will be treated in alphabetical order before proceeding to the Pea Family.

Barberry *(Berberis)*

The barberries of southern California are a difficult group in which to separate the species. Although as many as eight species have been recognized for our area in recent treatments, their delimitation and relationships are extremely problematical. For our purposes it is convenient to regard them as belonging to three groups.

Shinyleaf Barberry *(Berberis pinnata)*

The most distinct of our species, Shinyleaf Barberry,

occurs in the outer Coast Ranges of Santa Barbara, Los Angeles, and San Diego counties. It is an erect shrub about 3 feet tall which grows in shaded canyons and flowers mostly in late winter. Its leaves, subdivided into 5 to 9 leaflets (sometimes more), are about 2 to 5 inches long. The broad individual leaflets are about 1 to 2 inches long and more than half as wide. The characteristic leaves are bright glossy green on both surfaces. The others all have dull or gray-green leaves, at least below. The bright yellow flowers of Shinyleaf Barberry are grouped in fairly large clusters. The berries, which are about ¼ inch long, are blue and covered with a waxy bloom.

California Barberry *(Berberis dictyota)*

What I have grouped here is a complex consisting of several closely related and intergrading species. Unlike Shinyleaf Barberry, they occur on dry brush-covered hills throughout the Coast Ranges of southern California. The leaves are simliar in size to those of the preceding, but are thicker and dull gray-green. The flowers and fruits are likewise similar, and occur in large groups. California Barberry flowers in spring.

Nevin Barberry *(Berberis nevinii)*

Here I am treating another complex of species in which the relationships are very poorly understood. These plants have only 5 to 9 flowers in each cluster, and their leaves are much narrower, the leaflets being rarely over ½ inch wide. The leaves are dull gray-green. Some of the populations in this collective species have red berries; others are blue or even dull brown at maturity. In some, the outer flesh of the berry is inflated and dry, with a pocket of air inside; in others it is solid. A proper understanding of this horticulturally useful group will have to await a careful comparison of bushes from different populations growing side by side. These plants flower in spring, and scattered populations occur from the dry slopes of the San Gabriel Mountains and San Fernando Valley through the drier parts of Riverside and San Diego counties and locally across the Mojave Desert.

Southern Mountain Misery *(Chamaebatia australis)*

Erect sticky shrub with very finely divided leaves, the leaflets finely subdivided. The shrubs are mostly about 3 feet or more in height, and the white, regular, strawberry-like flowers appear in winter and spring in small clusters on the ends of the branches. They are about ¾ inch across. Southern Mountain Misery occurs in California only in southern San Diego County, from the Potrero Grade to the San Miguel and Otay mountains.

Chaparral Flowering Ash *(Fraxinus dipetala)*

This attractive shrub is about 6 to 15 feet tall. The leaves are several inches long and divided into 3 to 9 leaflets. The white flowers are about ¼ inch long, but clustered and attractive, and, distinctively, have only 2 stamens and 2 petals. The fruits are also distinctive, being flattened nuts with a wing approximately 1 inch long extending from them. The flowers of this deciduous ash appear in spring with the young leaves. Chaparral Flowering Ash occurs from the Santa Ana Mountains northward to the Sierra Nevada and Coast Ranges, being found through most of southern California away from the deserts.

Bladderpod *(Isomeris arborea)*

A glance at the plant will serve immediately to distinguish this common fast-growing shrub from all other members of the native flora. Its characteristic inflated pods 1 to 2 inches long, narrowed to a stalk at the base (but above the calyx), divided leaves, and yellow flowers make it unmistakable. Bladderpod exudes a very strong rank odor when the leaves are crushed. Flowers and pods may usually be found on it at any time of year, but are most frequent in later winter and spring. It occurs along the coast from the Santa Monica Mountains south and in the dry interior ranges from Santa Barbara County southward and eastward to the deserts.

Southern California Black Walnut
(Juglans californica)

Our native walnut is a common large shrub or small tree of brushy hillsides from Santa Barbara south to the Santa Ana and San Bernardino mountains. It is uncommon near the margins of its range, north of Santa Barbara and south of Los Angeles. The staminate flowers are borne in long tassel-like clusters, the pistillate ones in small groups; both are greenish and inconspicuous. Like other walnuts, it is deciduous, and the flowers appear in late winter with the young leaves. Mature leaves are about 6 to 12 inches long, and are divided into 9 to 18 leaflets 1½ to 3 inches long. The fruits are typical walnuts, about 1 inch long. When young, they are covered with a fleshy, fibrous, green, smooth covering which later dries up, turns black, and splits off, exposing the wrinkled brown shell.

Creosote Bush *(Larrea tridentata)*

Creosote Bush covers hundreds of square miles in the deserts of southern California. Its shiny, dark green, opposite leaves are strong-scented and divided into 2 leaflets. The brittle branches are jointed, with dark rings at the point of attachment of the leaves. The attractive flowers appear in winter and spring. They are ½ inch or more across, and have 5 bright yellow petals. The fruits are also fairly conspicuous, and the bushes often fruit heavily. The fruits are rounded, up to about ¼ inch long, and densely covered with white hairs. Creosote Bush, an erect shrub, usually 4 to 8 feet tall, is generally recognizable even at a distance by its dark green color and dark stems. In seasons of drought,

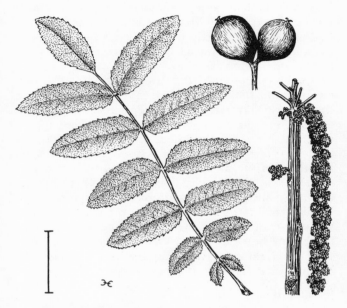

Southern California Black Walnut

however, the shrubs become paler or yellowish green. Normally they contrast very markedly with the predominantly gray-green desert shrubs belonging to other species. Often the bushes growing along roadside shoulders are darker green than those growing out in the desert, as they receive a better supply of water washed off the road. This is one of a very interesting group of plants that occur both in the arid regions of South America and in those of North America, being entirely absent from the more than 5,000 miles of moist tropics in between. All other species of *Larrea,* including one very similar to our North American plant, are restricted to South America, centering in northern Argentina.

Poison Oak *(Rhus diversiloba)*

It is important to become familiar with this common shrub, since contact with it causes a serious inflammation of the skin in many people. It can be recognized by reference to plate 1, which shows its 3-parted leaves, small greenish flowers, and smooth white or brown berry-like fruits that are about ¼ inch in diameter. Its leaflets are about 1 to 3 inches long, with characteristically wavy margins. Poison Oak may stand alone as an erect shrub, grow up among others, sprawl over rocky or grassy hillsides, or twine, vine-like, up tall trees, to which it attaches itself by aerial roots like those of ivy. It is deciduous; when it has lost all its leaves in the winter, it is difficult to recognize. This may be particularly disastrous, as particles of its oil carried in smoke are sufficient to cause the rash, and in a leafless state it is readily picked up with other brush and burned. Poison Oak is found throughout southern California except in the deserts and high mountains. It flowers in spring.

Squaw Bush *(Rhus trilobata)*

Although Squaw Bush might superficially be confused with its noxious relative, Poison Oak, it has none of the allergy-producing properties of that species. Its leaves also are mostly divided into 3 leaflets, but sometimes the terminal one is deeply notched so that the leaf appears to be divided into 5 instead of 3 parts. The terminal leaflet is rarely more than 1 inch long and lacks a distinct stalk, which is always present in Poison Oak. Squaw Bush is deciduous, and its yellowish clustered flowers appear in early spring before the first leaves. Its fruits are flattened, reddish, covered with short sticky hairs, and up to ¼ inch across. From them the Indians used to prepare a refreshing, lemonade-like drink. Squaw Bush occurs on brushy slopes and along streams throughout southern California, including the desert mountains.

Wild Rose *(Rosa)*

Although it is a simple matter to recognize a rose, it is not easy to tell our various kinds apart. All the wild species have single flowers, and they are very fragrant. With their 5 pinkish petals, they are among our most attractive wildflowers; and in autumn and winter their bright red berries are conspicuous along our roadsides. There are three native roses in southern California.

Wood Rose *(Rosa gymnocarpa)*

Wood Rose is a slender delicate rose of shaded places that flowers in spring. In southern California it occurs only in the Palomar Mountains of San Diego County; it can be distinguished from all other species by the fact that its ripe fruit has no sepals attached.

California Rose *(Rosa californica)*

California Rose is common along roadsides and streams throughout the coastward ranges of southern California. Its pink flowers are 1 to 2½ inches across. The shrubs are usually about 4 to 8 feet tall. The prickles on its stems are flattened, stout, and recurved. California Rose blooms in spring and summer.

Mountain Rose *(Rosa woodsii)*

Mountain Rose is similar to the preceding, but is armed only with straight, weak prickles. It occurs mostly above 4,000 feet elevation, from the vicinity of Mount Pinos to the San Gabriel and San Bernardino mountains. It flowers in spring and summer.

Wild Blackberry *(Rubus ursinus)*

The prickly, sprawling stems of Wild Blackberry are often about 3 to 6 feet long. The flowers, with their white petals and numerous stamens, are similar to those of a rose, and are about 1 inch across. Each leaf is deeply 3-lobed or divided into 3 separate leaflets, and the leaflets themselves range from about 1 inch to more than 3 inches long. The succulent berries (which are actually compound fruits composed of a number of smaller fleshy berries corresponding with the bumps) are about ½ inch long, at first green, and shiny black at maturity. They are edible and tasty but difficult to gather in quantity. Wild Blackberry occurs throughout southern California except for the deserts and high mountains, and is found most frequently in shaded places.

Wild Blackberry

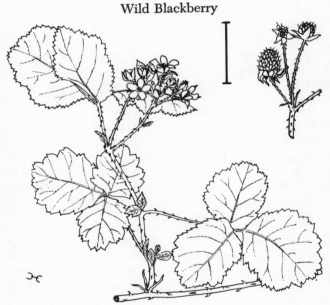

Himalaya Berry *(Rubus procerus)*

The Himalaya Berry is cultivated and sometimes established in waste places. It can be distinguished by its 5 leaflets, all attached to a single point at the top of the leaf stalk. Both species flower in late winter or spring and fruit in summer, Wild Blackberry somewhat earlier.

[32]

Western Raspberry *(Rubus leucodermis)*

Western Raspberry is similar in general aspect to Wild Blackberry, but the individual stems are stouter and whitish, being thinly covered with a waxy substance when young. Its leaves are divided into either 3 or 5 leaflets. The flowers are less than ½ inch across. As in all raspberries, the flesh of the fruit separates neatly from the stalk as a single unit, whereas in blackberries the flesh is more or less fused with the stalk and difficult to break off. Western Raspberry flowers in early summer and bears its black, purplish, or reddish fruits later. It is found in the pine belt of the region about Mount Pinos and in the San Gabriel, San Bernardino, Palomar, and Cuyamaca mountains.

Blue Elderberry *(Sambucus caerulea)*

Although elderberry is a member of the Honeysuckle Family, it seems to have little in common with the familiar garden vine. Its large leaves are divided into 3 to 9 leaflets, each usually 1 to 4 inches long and finely toothed along the margins. The tiny white flowers of elderberry are densely packed in flat-topped clusters several inches across, followed by small tart blue berries. Blue Elderberry is common in the Coast Ranges of southern California and rare in the piñon pine—juniper belt of the mountains of the Mojave Desert. It is a large shrub or small tree. In the higher San Bernardino Mountains grows a related species, *Sambucus microbotrys,* which is a rounded shrub rarely more than 3 feet tall, with rounded flower clusters and bright red berries.

THE PEA FAMILY

All members of the Pea Family can be recognized by their fruits, exemplified by the familiar pods of peas, beans, or sweet peas. These pods split along two lines, above and below, and usually contain a number

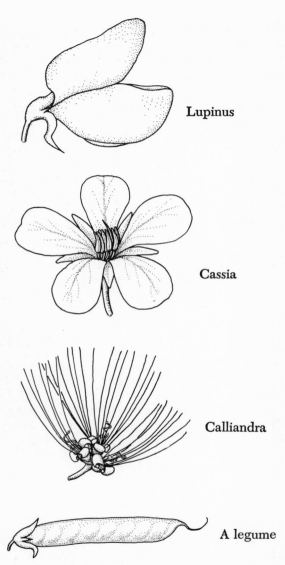

Lupinus

Cassia

Calliandra

A legume

Pea Family: Flowers and a fruit

[34]

of seeds. On hot summer days the pods may split explosively with a popping sound. Many members of this group have a highly irregular flower like that of the sweet pea, but others do not; all have 5 petals. Most genera have compound leaves, but the few that have simple leaves are treated here with the others for convenience.

In five of our genera the flowers are more or less regular in outline, and the petals are nearly equal in size and shape. In *Cassia,* which has conspicuous yellow petals, the leaves are only once divided, whereas in the other four genera the primary leaflets are again subdivided into leaflets. Three of these have tiny, inconspicuous petals and conspicuous bunches of stamens. *Acacia* has numerous short, curved spines like a cat's claw; *Prosopis* (mesquite) has longer straight paired or single spines; *Calliandra* is without spines and has large bright red flowers. *Hoffmanseggia* differs from these three in having conspicuous yellow petals.

Our remaining five genera have highly irregular pea-like flowers. In *Lupinus* (lupine) the leaflets arise from a single point at the top of the leaf stalk. *Amorpha* is unique among southern California representatives of the Pea Family in having only a single petal, but in all other respects is a typical pea. *Cercis* has glossy undivided leaves that are heart-shaped at the base. *Lotus* has yellow flowers. *Pickeringia* is a spiny shrub of the chaparral of the more coastward mountains; *Dalea* is confined to the desert.

Cat's Claw *(Acacia greggii)*

Spreading shrub about 3 to 6 feet tall, sometimes taller, found throughout the deserts. The branches are armed with stout, short, curved spines which give Cat's Claw its English name. The leaves are about 1 to 2 inches long, each divided into 1 to 3 pairs of leaflets, these further subdivided into 4 to 6 pairs of leaflets. The flowers are yellow and crowded in spikes about 1 inch long. The pods are light brown, constricted between the seeds, and mostly 2 to 5 inches long. Scattered colonies of Cat's Claw are sometimes found in southern California away from the desert.

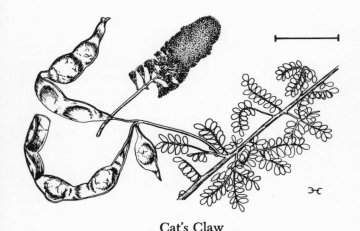

Cat's Claw

California False Indigo (*Amorpha californica*)

A large shrub of chaparral, oak woodland, or pine forest, found from the Santa Ana and Santa Rosa mountains north in the Coast Ranges. The leaves are mostly 4 to 8 inches long, subdivided into 11 to 27 leaflets up to 1 inch long. The flowers have a single purplish petal up to nearly ¼ inch long, and are crowded in upright spikes. They appear in late spring and early summer. The brown pods are about ¼ inch long and contain 1 or 2 seeds. California False Indigo has small, dark, prickle-like glands along the branchlets which are lacking in the very similar *Amorpha*

[36]

fruticosa, Desert False Indigo, found from the San Bernardino and San Jacinto mountains south into San Diego County.

Fairy Duster *(Calliandra eriophylla)*

Although rare, being found along washes in Imperial and easternmost San Diego counties, Fairy Duster is included here because it is one of our most attractive native shrubs. It has thick intricate dark branches (but no spines) and rarely exceeds 1 foot in height. The individual leaflets are further subdivided into smaller leaflets. The regular flowers have tiny, inconspicuous petals and very attractive bunches of red stamens nearly 1 inch long. It has brown flat pods about 2 inches long. Since the flowers are grouped, the stamens of the flowers are bunched into airy red clusters. Flowers appear in late winter.

Armed Senna *(Cassia armata)*

Armed Senna, although it is often 3 feet tall and found throughout our deserts, is rarely noticed except when it is in flower. For most of the year it is leafless, and seems to consist only of a cluster of greenish twigs, but in spring it bursts forth with a dense cover of attractive bright yellow flowers about 1 inch across and sparse green leaves 2 to 6 inches long with very scattered leaflets. Its flowers are regular and not at all pea-like. Armed Senna has rounded pods about 1 to 2 inches long.

Western Redbud *(Cercis occidentalis)*

Western Redbud is a large deciduous shrub of the pine belt on the desert slopes of the Laguna and Cuyamaca mountains in San Diego County. Its reddish-purple, generally pea-like flowers and rounded, entire glossy leaves, 1 to 3 inches across and heart-like at the base, make it unmistakable. The flowers appear in late winter before the leaves.

Indigo Bush *(Dalea)*

There are more kinds of Indigo Bush among the southern California shrubs than any other group of the Pea Family. Collectively, they can be recognized by their bluish or purplish pea-like flowers and desert habitat. *Dalea spinosa* and *D. schottii* have simple leaves, whereas the other four species have compound leaves: in *D. emoryi* and *D. polyadenia* the flowers are in dense, head-like clusters; in *D. arborescens* and *D. fremontii*, in loose spikes.

Mojave Dalea *(Dalea arborescens)*

A rather spiny shrub about 3 feet tall, the leaves divided into 3 to 7 leaflets and densely covered with short hairs, giving them a silvery appearance. The rather loose spikes of dark blue flowers appear in spring. Mojave Dalea is known only from the north and east region of Barstow on the Mojave Desert of San Bernardino County.

Emory Dalea *(Dalea emoryi)*

A very intricately branched shrub, often about 2 feet tall, common on the Colorado Desert. The branches are densely covered with a white felt-like pubescence, through which are sprinkled orange glands that stain the clothing or other material with which they come into contact. The rose-purplish flowers, which appear in early spring, are only about ¼ inch long, grouped in dense, flattened head-like clusters. The leaves are mostly 1 to 2 inches long, and usually divided into 5 to 7 leaflets; when crushed they emit a strong agreeable odor. Emory Dalea is of interest as the host of the rare parasitic *Pilostyles thurberi*. The parasite grows inside the branches of the host, and its purple flowers, about ⅜ inch long, emerge through the fissures in the bark. Occasional plants of Emory Dalea may be colored purple in spring if they are infested.

Fremont Dalea *(Dalea fremontii)*

Woody shrub 1 to 3 feet tall, the leaves smooth or only slightly hairy. The leaves are about 1 inch or a little more in length, and usually divided into 3 or 5 leaflets. The deep purple flowers appear in spring. Fremont Dalea occurs in washes and on slopes throughout the Mojave Desert, whereas the closely similar California Dalea, *Dalea californica,* occurs along the western margins of the Colorado Desert.

Nevada Dalea *(Dalea polyadenia)*

Similar to Fremont Dalea, but the leaves are usually divided into 7 to 11 leaflets, and the entire plant is densely covered with conspicuous dot-like glands. Nevada Dalea has densely clustered violet-purple flowers; in Fremont Dalea they are loosely arranged in a spike. Nevada Dalea occurs in scattered localities on the Mojave Desert.

Mesa Dalea *(Dalea schottii)*

Mesa Dalea is an intricately branched spiny shrub that is often about 5 feet tall: The leaves are narrow and undivided, but the bluish or purplish flowers are like those of the other Daleas. They appear in late winter. Mesa Dalea occurs widely in the washes and on the slopes of the Colorado Desert, and is particularly common along its western margins.

Smoke Tree *(Dalea spinosa)*

The familiar Smoke Tree is normally leafless, but its intricate grayish branches make it appear like a puff of smoke when viewed at a distance. The undivided leaves are found on seedlings and occasionally at other times of the year, and the attractive loose clusters of bright bluish-purple flowers are abundantly borne in early summer. Smoke Tree is found in sandy washes throughout the Colorado Desert, and on the southern Mojave Desert from Daggett east.

Rushpea *(Hoffmanseggia microphylla)*

Rushpea is a shrub with numerous slender rush-like greenish branches arising from the base, usually about 3 feet tall. The scattered leaves are divided into 3 leaflets and then the leaflets are further subdivided, the entire leaf being about 1 inch long. The dull yellow flowers are regular and less than ½ inch across, with 5 conspicuous petals. The flat brown pod is about 1 inch long. Rushpea occurs on the Colorado Desert and flowers in late winter and early spring.

Deerweed *(Lotus scoparius)*

Rounded shrub, often about 2 to 3 feet tall, with slender green branchlets. Deerweed is only slightly woody at the base. Its yellow flowers, sometimes tinged with reddish, are about 1/3 inch long and occur in small bunches along the stems. They are most common in spring and summer but may be found in almost any month of the year. The pods are about ½ inch long, hard, and usually contain 2 seeds. Deerweed is found throughout southern California up to the margins of the desert. A number of other species of *Lotus* are sometimes slightly woody at the base, including Desert Rock-pea, *L. rigidus,* which has yellow flowers about ½ inch or more long and occurs on or near the deserts.

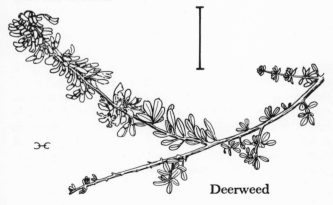

Deerweed

Tree Lupine *(Lupinus arboreus)*

Lupines have very distinctive leaves, with several narrow leaflets arising from a single point at the top of the stalk. Tree Lupine is a yellow-flowered species which grows on the coastal dunes and bluffs from the vicinity of Ventura northward and can scarcely be confused with anything else. All other shrubby species of southern California have blue or purplish flowers. Tree Lupine is a true shrub, ranging from 2 to 6 feet in height. It flowers in spring.

Pauma Lupine *(Lupinus longifolius)*

This is perhaps the commonest of a group of three bluish-flowered shrubby lupines found in southern California. Its leaves have 6 to 9 leaflets which are about 1 to 2½ inches in length, covered on both sides with long hairs. Pauma Lupine ranges from Ventura County southward in the coastward ranges. North of this area it is replaced by the very similar and scarcely distinct Silver Lupine, *Lupinus albifrons,* whereas on the immediate coastal dunes and bluffs from Los Angeles County north occurs Dune Lupine, *L. chamissonis.* The separation of this trio of closely related species is a vexing job even for the professional, and they can best be named by reference to the locality where they were collected. In general, however, Silver Lupine tends to have the leaflets more densely and silvery hairy than do the others. All three species bloom primarily in spring. A number of the other perennial lupines in southern California may occasionally be somewhat woody at the base.

Chaparral Pea *(Pickeringia montana)*

Chaparral Pea is a spiny evergreen shrub 2 to 6 feet tall, found on brush-covered slopes. Its leaves are usually 3-parted, the leaflets ¼ to ½ inch long, or more rarely undivided. The rose-purple flowers, which appear in late spring, are about ¾ inch long. Technically

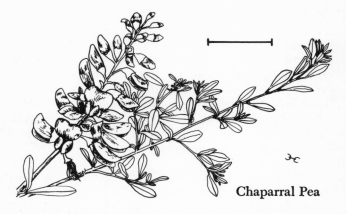

Chaparral Pea

it can be distinguished from all other shrubs of our area with pea-like flowers by the fact that its stamens are not united, but each is free to the base. Chaparral Pea very rarely sets seeds, propagating mostly by underground runners. In southern California it is highly colonial, occurring scattered throughout the Coast Ranges, but not at low elevations.

Mesquite *(Prosopis glandulosa)*

Mesquite is a deciduous large shrub or small tree, widely distributed in the deserts and interior semiarid regions of southern California. Like *Acacia* and *Calliandra*, it has tiny petals, and the conspicuous part of its yellowish flowers consists of the stamens. The flowers are tightly clustered in slender spikes mostly 2 to 3 inches long, borne in spring. The pods range from 3 to 8 inches long and are somewhat constricted between the seeds. They ripen in fall and are staple food for various Indian tribes, being gathered, ground into meal, and made into a kind of bread. The branches of Mesquite are covered with scattered spines about ½ inch long. Its leaves are usually divided into 2 primary leaflets, each of which is then subdivided further into 7 to 18 pairs of leaflets.

Screwbean *(Prosopis pubescens)*

Screwbean is similar to Mesquite, but has shorter spines and fewer pairs of leaflets on the primary divisions of the leaf. Its most distinctive feature consists of its tightly coiled pods, 1 to 1½ inches long, which are borne in clusters of 2 to 10 and ripen in summer. Screwbean is found in scattered localities on the deserts and at the southern end of the San Joaquin Valley. It flowers in early summer.

GROUP VII

Flowers clustered into dense heads, yellow, white, blue, or purple, these heads surrounded by a circle of more or less overlapping bracts.

In this group is included the Sunflower or Daisy Family. The members of this large family will be treated in three groups: VIIA, including those in which the marginal flowers of the head are modified into petal-like rays, thus being very different from the central flowers of the head; VIIB, consisting of plants that lack rays but have bright yellow flowers; and VIIC, consisting of plants that lack rays and have whitish, pale yellow, or purplish flowers.

GROUP VIIA

Members of the Sunflower Family that have the marginal flowers modified into distinct rays.

In this group, species of *Coreopsis, Encelia,* and *Viguiera* have a small erect scale at the base of each flower in the head; the others lack them. In *Eriophyllum* and *Senecio* the leaves are deeply divided; in the others they are not. *Venegasia* has broad soft leaves 2 to 7 inches long. *Gutierrezia microphala* has only 2 to 4 flowers per head, fewer than in any other member of the group. *Haplopappus cuneatus* has broad hard leaves ¼ to ¾ inch long; all others have narrow leaves;

H. linearifolius has 11 to 18 conspicuous ray flowers, the others fewer than 10. Remaining species are *Haplopappus* or *Gutierrezia sarothrae*.

Giant Sea Dahlia *(Coreopsis maritima)*

Giant Sea Dahlia can easily be recognized by its stout fleshy trunks, often about 3 feet tall, surmounted by a tuft of large, well-dissected leaves that dry up during summer. The bright yellow heads are 2 to 3 inches across, and are at their best in late winter. Giant Sea Dahlia occurs in coastal bluffs above the ocean from the vicinity of Point Dume northward in scattered localities to southern San Luis Obispo County. It is common also on most of the offshore islands.

California Encelia *(Encelia californica)*

A softly woody shrub that sprawls and attains about 4 feet in height. The leaves are broad and mostly about 1½ to 2½ inches long. The heads are about 2 inches across, with yellow rays and purplish-brown centers, and appear mostly in spring. It ranges through the coastal mountains from the vicinity of Santa Barbara south, and also extends inland to the area of Riverside, where it hybridizes with the following:

Desert Encelia *(Encelia farinosa)*

Desert Encelia is a rounded, distinctly woody shrub, often about 2 feet tall. The heads, which appear in spring, are somewhat smaller than those of California Encelia, and the central flowers are usually yellow like the rays. The leaves are silvery. Desert Encelia is common throughout the deserts, and extends west to the region of Lake Elsinore.

Virginia City Encelia *(Encelia virginensis)*

Similar to Desert Encelia, but with smaller heads borne singly at the tops of the leafy branches. In Desert Encelia the heads are grouped in open clusters at the top of each leafy stem. Virginia City Encelia flowers in spring and occurs on the Mojave Desert.

[44]

Golden Yarrow *(Eriophyllum confertiflorum)*

A very common slightly woody shrub that ranges up to about 2 feet in height. The leaves are deeply divided and more or less covered with a loose gray pubescence. The small yellow heads are densely grouped in flat-topped clusters. Golden Yarrow flowers in late spring and early summer and occurs throughout southern California except in the deserts.

Sticky Snakeweed *(Gutierrezia microcephala)*

This low shrub of the Mojave Desert has linear leaves and can readily be distinguished by the fact that its yellow heads have only 1 or 2 rays and only 1 or 2 disc flowers. Sticky Snakeweed blooms in late summer and fall.

Broom Snakeweed *(Gutierrezia sarothrae)*

Broom Snakeweed and its close relative, *Gutierrezia bracteata,* occur together throughout the Coast Ranges and also in the desert mountains. Their yellow heads are narrow, less than ¼ inch high, and have from 3 to 7 relatively short rays. Technically they can be distinguished from all species of *Haplopappus* by the fact that their seeds have only short scales at the summit instead of the crown of bristles evident in the latter. Broom Snakeweed blooms in fall, for the most part, although occasional plants may be found flowering at other times of the year.

Cooper Goldenbush *(Haplopappus cooperi)*

A common low shrub of the Mojave Desert which bears its golden-yellow flowers in spring. The heads have only 1 or 2 short rays, and 5 to 9 central flowers. The very narrow leaves are about ½ inch long.

Wedgeleaf Goldenbush *(Haplopappus cuneatus)*

A low densely branched woody shrub of rock crevices that can easily be recognized by its broad leathery leaves which are about ½ inch long. There are from

Goldenbush *(Haplopappus)*

H. linearifolius   *H. cooperi*   *H. venetus*

1 to 5 rays. Wedgeleaf Goldenbush flowers in fall and winter, and occurs on rocky slopes in the mountains throughout our area.

Heather Goldenbush *(Haplopappus ericoides)*

One of a group of three species of goldenbush with needle-like, short, densely set leaves. All three flower primarily in the fall. The present species is confined to sandy flats in the immediate vicinity of the ocean, and has leaves less than ½ inch long. In the other two, *Haplopappus pinifolius,* Pine Goldenbush, and *H. palmeri,* Palmer Goldenbush, the leaves are ½ inch to 1½ inches long. Pine Goldenbush occurs on dry slopes away from the coast from northern Los Angeles

County south, and has very sparsely hairy seeds. Palmer Goldenbush ranges from southern Ventura County south, and has seeds densely covered with hairs.

Narrowleaf Goldenbush *(Haplopappus linearifolius)*
In this common shrub the bright yellow heads are about 1½ inches across and have 11 to 18 conspicuous rays. The narrow leaves are about 1 inch long or a little longer. Also very distinctive is its blooming period, late winter or earliest spring. Narrowleaf Goldenbush occurs nearly throughout southern California, extending east to the margins of the deserts.

Bush Groundsel *(Senecio douglasii)*
A common bushy perennial, similar in general appearance to the preceding, but with the leaves divided into linear segments and covered with short gray hairs. Bush Groundsel flowers in summer and fall and occurs throughout southern California except in the deserts.

Canyon Sunflower *(Venegasia carpesioides)*
Canyon Sunflower is scarcely woody at the base, but, as it often ranges up to 5 feet in height, it is included here. Its leaves are soft, dark green, and from 2 to 7 inches long and 1 to 4 inches across. Canyon Sunflower blooms in spring. It does not range far from the coast, but occurs on shaded slopes and burns, being less common near the northern and southern limits of our area.

Viguiera Deltoides
Our two common species of *Viguiera* can be distinguished from *Encelia* by the lobed margins of their leaves. Both bloom in spring primarily, and both have yellow rays and central flowers. The present species has heads about 1½ inches across and egg-shaped leaves. It occurs on the deserts.

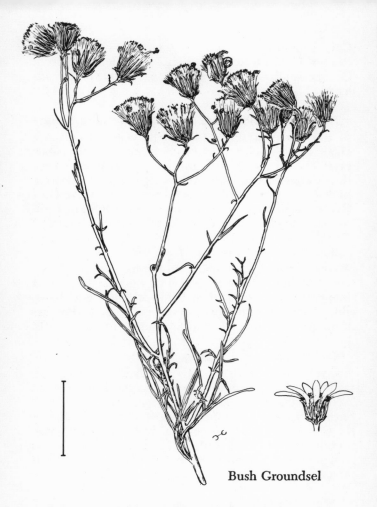

Bush Groundsel

San Diego Sunflower *(Viguiera laciniata)*

San Diego Sunflower has narrower leaves than *Viguiera deltoidea*. It is densely leafy, with crinkled, lobed leaves 1 to 2 inches long. This species is very common in the vicinity of San Diego and elsewhere in the southwestern part of San Diego County.

[48]

Members of the Sunflower Family that have bright yellow flowers but lack rays; all the flowers within a given head are thus of one kind.

Among this group, *Trixis* has deeply 2-lipped flowers; *Encelia* and *Bebbia* have a small scale at the base of each flower; *Acamtopappus* has a fringe of hairs around the margin of each bract on the head; *Lepidospartum* has reduced, scale-like leaves. In *Chrysothamnus* and *Haplopappus* the bracts of the head overlap one another, shingle-like; in the former, the bracts are in distinct overlapping rows; in the latter they are not. In *Peucephyllum* the bracts are in two distinct series; in *Tetradymia*, in only one series.

Goldenhead (*Acamtopappus sphaerocephalus*)

Goldenhead is a common low rounded desert shrub with nearly globose golden-yellow heads about ¼ inch thick. Goldenhead occurs throughout the Colorado and Mojave deserts. Its most distinctive characteristic is the fringe of hairs bordering each of the bracts on its heads. Goldenhead flowers in spring.

Rush Bebbia (*Bebbia juncea*)

Slender shrub with ascending green branches, often nearly leafless at flowering time (late spring and early summer). Rush Bebbia is found throughout our deserts. Its leaves are narrow and mostly less than 1 inch long. It can easily be distinguished by the scales subtending each of its golden-yellow flowers and by the characteristics given above.

Rabbitbrush (*Chrysothamnus*)

As a group, the rabbitbrushes can be distinguished by their narrow, rayless, golden-yellow heads with the bracts lined up in clear-cut vertical rows. They flower mostly in fall. In two of our species the leaves are needle-like and nearly round in cross section: *Chrysothamnus teretifolius* and *C. paniculatus*. In the latter the heads tend to be in more or less rounded, open clusters, which in the former are narrower. Both species are relatively rare, and occur along the margins of the deserts. *Chrysothamnus viscidiflorus* occurs in the Santa Rosa, San Jacinto, San Bernardino, and San Gabriel mountains and near Mount Pinos. The branches are smooth or nearly so. The commonest species of rabbitbrush in southern California is treated in a separate paragraph.

Rabbitbrush *(Chrysothamnus nauseosus)*

An erect shrub, often sparsely leafy at flowering time, with broom-like branches densely covered with short matted gray-green or whitish hairs. The narrow leaves range from ¾ inch to 2 inches in length. The golden-yellow heads are aggregated into more or less flat-topped clusters, and the plants are often quite showy when in bloom. This rabbitbrush is common over much of the Mojave Desert and bordering mountains.

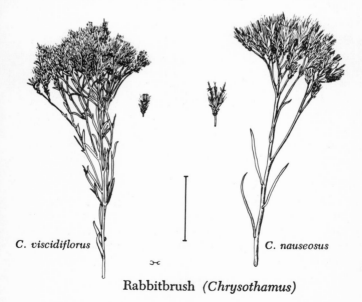

C. viscidiflorus   *C. nauseosus*

Rabbitbrush *(Chrysothamus)*

Rayless Encelia (*Encelia frutescens*)

Rayless Encelia, found throughout our deserts, is a rounded shrub usually about 2 to 3 feet tall. Its leaves are mostly ½ to 1 inch long and about half as broad. Each leafy branch is surmounted by a single yellow flowering head about ½ inch across. Rayless Encelia flowers in spring.

Scalebroom (*Lepidospartum squamatum*)

The rounded gray-hairy leaves of Scalebroom are rarely seen, for the plant is generally found in a leafless condition. Scalebroom is a rounded shrub 3 to 6 feet tall, with rigid, broom-like green branches. Its golden-yellow heads, which are frequent from summer into the rainy season, are about ¼ inch thick and surrounded by blunt, overlapping bracts. Scalebroom is common on dry slopes and especially along washes throughout much of southern California away from the immediate coast, but is only occasional on the desert. In the region of Whitewater Wash above Palm Springs it is abundant on sandy flats.

Pigmy Cedar (*Peucephyllum schottii*)

Pigmy Cedar is the only desert-dwelling member of the Sunflower Family with dense-set, needle-like leaves. These leaves range up to 1 inch in length. The shrubs are up to about 8 feet tall, and have a dark green appearance at all seasons of the year regardless of drought. The yellow heads are up to ½ inch long and rather stout. Pigmy Cedar is most frequent in rocky places such as the heads of canyons. It blooms in late winter and spring.

Horsebrush (*Tetradymia*)

Several rather similar species of shrubs belonging to this genus occur in southern California, where they bloom in late spring and summer. All have golden-yellow flowers in heads about ½ inch high and narrow clustered leaves.

Tetradymia glabrata has 4 flowers per head and is not spiny. Its mature leaves are smooth or nearly so. It occurs on the western Mojave Desert. *T. canescens* is similar, but its mature leaves are covered with short gray hairs, and it is found mostly above 4,000 feet in the San Bernardino and San Gabriel mountains, north along the east side of the Sierra Nevada. *T. stenolepis* has 5 flowers in each head and is covered with spines; it is found on the Mojave Desert. The remaining two species have 6 to 9 flowers per head: *T. axillaris* is spiny and ranges from the Mojave Desert northward; *T. comosa* is not truly spiny, although it sometimes has spine-tipped leaves; it is most common along the western borders of the Mojave Desert.

California Trixis *(Trixis californica)*

Easily recognized by its deeply 2-lipped golden-yellow flowers. The dark green leaves have relatively smooth margins, and are about 1 inch long or somewhat longer, and about half as wide. The flowering heads are borne in ample clusters at the ends of the branches. California Trixis is a low rounded leafy shrub that blooms in late winter and is found in somewhat sheltered places on the Colorado and southern Mojave deserts.

GROUP VIIC

Plants of the Sunflower Family that have dull whitish, very pale yellowish or purplish flowers and lack rays.

Pluchea is the only one of the group that has bright purplish flowers. *Franseria* and *Hymenoclea* have two kinds of heads on each plant: staminate heads borne near the tops of the branches, in which each flower has a scale-like bract at its base; and pistillate heads below. In *Franseria* the pistillate heads have the bracts modified into a bur-like structure, whereas in *Hymenoclea* they have broad papery wings. *Artemisia* has bitter, aromatic, often divided leaves with the strong odor of sage. In *Hofmeisteria* the stalks of the leaves are mostly 1 to 2 inches long, and the leaves themselves are up to 1/3 inch long. The species of *Baccharis* and *Brickellia* will be treated below as a unit.

Coast Sagebrush *(Artemisia californica)*

Coast Sagebrush covers many thousands of acres on the lower slopes of the Coast Ranges and Tehachapi Mountains, and extends east to the borders of the desert. It is a low rounded shrub, mostly 2 to 3 feet tall, densely covered with very narrow leaves ¾ inch to 3 inches long. These grayish leaves are mostly divided into linear segments, much narrower than those of *A. tridentata*. The heads are inconspicuous, nodding, and crowded at the ends of the branches. Coast Sagebrush flowers from late summer into the rainy season.

Tall Sagebrush *(Artemisia palmeri)*

Tall Sagebrush is relatively common in the southwestern corner of San Diego County. It ranges up to 6 feet in height, and can be distinguished by its narrow 3- or 5-parted leaves. Its best technical character is the presence of a scale-like bract at the base of each flower in the head.

Bud Sagebrush *(Artemisia spinescens)*

Bud Sagebrush is a low shrub, rarely more than 8 inches tall, found widely over the western Mojave Desert. It is densely leafy, with 5- to 7-parted leaves about ¼ inch long, and armored with long, sharp spines formed from the old lateral branches. This species flowers in spring, and is common on dry, rocky, or sandy flats in its area.

Sagebrush *(Artemisia tridentata)*

This is the commonest of a group of closely related species, and is the Sagebrush that is so common over the desert flats in the intermountain states. In southern California this group can readily be distinguished by the silvery grayish-green, flattened, divided leaves mostly 1 inch or less long, which are not nearly so narrow as those of Coast Sagebrush. Large spongy insect galls are common on their branches, especially in summer. These species bloom in late summer and fall, and have clusters of inconspicuous heads at the tops of the leafy branches. *Artemisia tridentata* ranges up to 10 feet or even exceptionally as much as 25 feet in height. Like all members of the group, it is pungently aromatic. Its leaves are mostly more than three times as long as broad. In southern California it occurs widely in dry ranges away from the immediate coast, and extends to the margins of the desert, although it is not normally found on dry flats on the true desert. In the mountains of the eastern Mojave Desert and on the desert slopes of the San Bernardino Mountains is found the similar *A. arbuscula,* which has shorter, broader leaves. A third member of the group, Rothrock Sagebrush, *A. rothrockii,* occurs at high elevations in the San Bernardino Mountains; the branches have a tendency to sprout and form roots where they come in contact with the ground.

Baccharis and *Brickellia*

Although they are very different technically, these two common shrubby genera are best treated as a unit for our purposes. In *Baccharis* the heads consist strictly either of staminate flowers, which set no seed, or of pistillate ones, and these flowers are found strictly on different plants. In *Brickellia,* however, the flowers are both staminate and pistillate and all plants are capable of setting seed. Further, in *Baccharis* the imbricated bracts around the head are green and leaflike in texture, whereas in *Brickellia* they are membranous.

In one group of these species belonging to the genus *Brickellia,* the heads are solitary, at least ½ inch long, on a distinct stalk at the top of each leafy branch. White Brickellbush, *Brickellia incana,* which flowers in late spring, is the only one of these that is densely covered with matted white hairs. It occurs on the western border of the Colorado Desert. Its leaves are broad and often 1 inch long. The other three species are found on rocky slopes in the desert mountains, and bloom in spring. One, *B. frutescens,* Rigid Brickellbush, has somewhat spiny branches and leaves less than ½ inch long. It is occasional on desert flats from the northern Colorado Desert northward. *B. arguta,* Pungent Brickellbush, which is widespread, has leaves egg-shaped in outline and usually sharply toothed; in *B. oblongifolia,* Mojave Brickellbush, the leaves are narrow.

Of the remaining species, only two have the leaves truncate to heart-shaped at the base. The most common of these is *B. californica,* California Brickellbush, which blooms in the fall and is found on dry slopes and washes throughout the mountains of southern California, although it is very rare on the desert. This spreading shrub ranges up to about 3 feet in height and has broad rounded leaves ½ inch to 2 inches long, with minute teeth around their margins. They are more or less densely covered with short gray hairs. The heads of this species are about ½ inch long, and the flowers exude a very fragrant odor in the evening. The leaves of California Brickellbush have short but distinct stalks, whereas those of its close relative, Nevin Brickellbush, *B. nevinii,* are attached directly to the stem. These leaves are also smaller and more densely white-hairy. Nevin Brickellbush blooms in the fall and is found in openings on brushy slopes from the San Gabriel and Santa Monica mountains north to the Santa Ynez Range and San Emigdio Canyon in southwestern Kern County.

All species not treated thus far belong to the genus *Baccharis* and hence have the characteristics listed

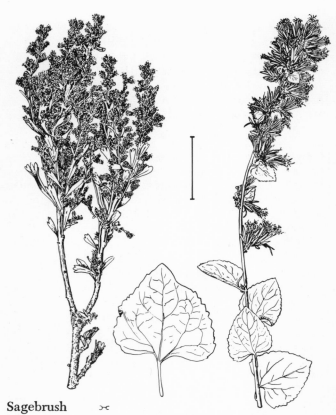

Sagebrush   ✂

California Brickellbush

above. In *B. plummerae,* found on shaded mountain slopes near the coast from the Santa Monica Mountains to the Santa Ynez Range, the leaves are elongate, acutely toothed, and covered with spreading sticky hairs. All species of *Baccharis* bloom primarily in the fall, and all the others have smooth, not hairy, leaves.

Broom Baccharis *(Baccharis sarothroides)*
A stout shrub 2 to 6 feet tall, nearly leafless at time of flowering, and broom-like with green branches. Broom

Baccharis occurs in sandy washes from the vicinity of San Diego, where it is frequent, eastward in scattered localities across the Colorado Desert. Its short leaves are narrowly linear, whereas in *B. sergiloides,* Squaw Waterweed, which is very similar in habit, the leaves are broader in their upper half than they are near the base. Squaw Waterweed ranges widely across the deserts and into the interior ranges in San Diego County.

Mule Fat *(Baccharis viminea)*

Mule Fat is a very common shrub which ranges up to 10 feet tall. Its willow-like leaves range from 1 to 3½ inches in length, and their margins are either uncut or evenly toothed all around. The small clusters of heads are borne on numerous short lateral branches. *B. glutinosa,* Seep Willow, is similar, but the leaves more often have toothed margins, and the heads are in a large cluster at the top of the stem. Mule Fat is found throughout southern California away from the deserts, whereas Seep Willow is found in wet places on the deserts and in the interior ranges. Both might be confused with *B. douglasii,* which is nearly herbaceous to the base but often 6 feet tall. It occurs in the Coast Ranges in moist places, as behind salt marshes on the immediate coast, and blooms in the late summer and fall; the two truly woody species bloom in the spring (although occasional flowers can be found on the bushes at any time of year).

Coyote Brush *(Baccharis pilularis)*

Coyote Brush is a common shrub that blooms in the fall and into the rainy season and occurs on open grassy hillsides throughout the outer Coast Ranges of our area. It is densely leafy, the leaves mostly about ½ inch to 2 inches long and ¼ to ½ inch wide, usually with about 5 to 9 coarse teeth in their upper part. They are narrow at the base and broadest above the middle. The heads are about ¼ inch high, and the bristles on the seeds of the pistillate plants form conspicuous white tufts on them when they are in fruit. *B. emoryi* has leaves that are somewhat similar in outline to those of Coyote Brush, but thinner; unlike that species, its flowers are borne on nearly leafless branches above the foliage. It is found from the southern end of the San Joaquin Valley south in the Coast Ranges, and always grows in wet places. Coyote Brush, on the other hand, is found on dry slopes.

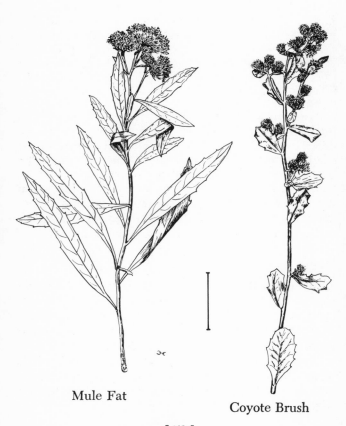

Mule Fat

Coyote Brush

Burrobrush *(Franseria dumosa)*

This low rounded grayish-green shrub is probably, next to Creosote Bush, the commonest shrub over most of the vast sandy tracts of both our deserts. It is intricately branched, with its white stems interwoven and becoming like spines in old age. Its clustered leaves are well divided and up to 1 inch long, and densely covered with short gray hairs. Its seeds are borne in spiny burs formed by the fusion of the bracts of the pistillate flowers, borne below the tassels of staminate flowers. Burrobrush flowers primarily in spring, but flowers may be found on bushes of this species after rains at almost any time of the year. A related species with undivided broad leaves densely covered with short white hairs beneath, *Franseria chenopodiifolia,* is common on Otay Mesa in southwestern San Diego County.

Arrowleaf *(Hofmeisteria pluriseta)*

The whitish flowers of Arrowleaf are borne in heads about 1/3 inch high, but its leaves are its most distinctive feature. Their stalks are at least five times as long as the arrowhead-like blades, which range up to 1/3 inch long. Arrowleaf is a low woody shrub, often found in rock crevices, which occurs mostly in canyons scattered over both deserts. It flowers in spring.

White Burrobrush *(Hymenoclea salsola)*

A spreading shrub common over both the Colorado and Mojave deserts, White Burrobrush has thread-like leaves and grayish bark. Its green flowering heads are inconspicuous, but its seeds are borne in a fused structure about ¼ inch long, surrounded with whorls of broad membranous wings set at right angles. The plants are mostly about 3 to 4 feet tall and usually found in sandy washes. They flower in late winter and spring.

Arrowweed *(Pluchea sericea)*

The English name of this erect shrub, which ranges up to over 10 feet in height, is derived from the use of its straight branches by the Indians for arrows. It is the only truly shrubby purple-flowered member of the Sunflower Family in southern California, growing along rivers and in wet places throughout. The leaves are willow-like, but covered with a dense mat of silvery hairs, mostly ¾ inch to 2 inches long. The heads of purplish flowers are clustered in terminal bunches on the leafy branches. Arrowweed is particularly abundant in low ground along the Colorado River.

GROUP VIII

Petals present, united at least at the base to one another and often forming a tube.

Plants of the Sunflower Family also have their petals united into a tube, but the flowers are clustered into dense heads surrounded by bracts. These plants are treated as Group VII. *Galium,* which has small greenish or whitish petals united at the base and whorled leaves, is treated with Group IX. Excluding these, the present group consists of a number of genera of shrubs native to southern California. Those with opposite or whorled leaves are treated in Group VIIIA; those with alternate leaves, in Group VIIIB.

GROUP VIIIA

Leaves opposite or whorled.

Most of the members of this group have irregular flowers, the petals unequal in size and shape, but *Symphoricarpos* has regular flowers. *Salazaria* is a desert shrub with spiny branches, the fruits with a bladdery covering. *Salvia,* with blue, lavender, or white flowers, and *Beloperone,* with red flowers, have only 2 stamens; the remaining genera have 4 or 5. *Lonicera* has 5 with anthers, *Penstemon* has 5 but one is without an anther, and the others have 4 stamens. *Mimulus* has red to creamy yellow flowers and no mint odor, whereas the three remaining genera do. Of these, *Lepechinia* has large white open corollas more than 1 inch long; *Trichostema* has fuzzy purple flowers with long-exerted stamens; and *Hyptis* is a tall desert shrub with purplish flowers less than ¼ inch long.

California Beloperone *(Beloperone californica)*

Twiggy desert shrub with numerous erect greenish branches, rather densely covered all over with short erect hairs. The leaves are opposite and mostly less than 1 inch long. They drop off during the dry season, leaving the plants bare. The narrow red flowers are about 1 inch long, 2-lipped, and grouped in small clusters at the tops of the stems. They are visited by hummingbirds. California Beloperone, which flowers in late winter, is common on the Colorado Desert, where it is particularly frequent in sandy washes.

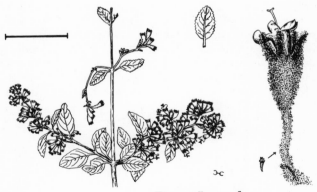

Desert Lavender

Desert Lavender *(Hyptis emoryi)*

Desert Lavender is a tall bushy shrub that is abundant in washes on the Colorado Desert and extends northward onto the southern Mojave Desert. It is very distinctive, with its minty odor and ashy-gray bark. The densely white-hairy leaves are about ½ inch across and rounded in outline, with shallowly toothed margins; they are borne in opposite pairs. The violet-blue flowers, less than ¼ inch long, are borne in dense clusters somewhat spaced along the upper ends of the branches. Desert Lavender flowers in spring.

Pitcher Sage   (*Lepechinia*)

The pitcher sages are low aromatic bushy shrubs with large hairy leaves about 3 inches long. The leaves are very aromatic when crushed, and are shallowly toothed all around the margins. Very characteristic of these plants, and the source of their English name, is the white corolla, which is over 1 inch long, ½ inch thick, and has an extended lower lip, thus resembling a china cream pitcher. The 4 black fleshy fruits are borne in the bottom of the thin papery calyx. Although four closely related species of pitcher sage are recognized in southern· California, they are separated on somewhat technical characters, and the distinctions between them are beyond the scope of this book. All flower in spring and early summer, and collectively they are found on brushy slopes below 4,000 feet in the coastal mountains, and also in the San Gabriel Mountains. They tend to be somewhat colonial and local in distribution.

Twinberry (*Lonicera involucrata*)

Twinberry, an erect shrub up to 10 feet tall, grows rather frequently in moist places along the immediate coast from the vincinity of Point Conception northward. Its dark green leaves are about 3 inches long and 1½ inches wide, and, like our other species of the honeysuckle genus, *Lonicera,* have entire margins. The yellowish flowers, tinged with red and about ½ inch long, are borne in pairs on a single stalk arising in a leaf-axil; the pair of flowers is surrounded by 2 small leaves which often become reddish when the fruit ripens. The corolla tubes are strongly swollen and distended on one side near the base. In early summer, when the pair of flowers has fallen, it is followed by 2 shiny black berries each about 1/3 inch in diameter.

California Honeysuckle *(Lonicera hispidula)*

California Honeysuckle is usually a clambering vine sprawling over other bushes or the ground, but it lacks tendrils. The leaves are rounded and usually about 2½ inches long and more than 1 inch wide; the upper pairs are fused together and encircle the stem, which passes through the fused unit. The purplish or pinkish irregular flowers are about ¾ inch long. California Honeysuckle flowers in spring and is occasional on relatively moist shaded slopes, often among dense brush and trees, from western Riverside and Los Angeles counties north in the Coast Ranges. The similar Chaparral Honeysuckle, *L. interrupta,* which also has the members of its upper pairs of leaves fused together, occupies approximately the same range, but is found on dry brushy slopes and is less likely to have a clambering habit. Its flowers are yellowish and smaller than those of California Honeysuckle.

Southern Honeysuckle *(Lonicera subspicata)*

On dry bushy slopes below about 5,000 feet throughout the coastal ranges of southern California, Southern Honeysuckle is found. It is usually an erect shrub, but sometimes sprawling. The leaves are rounded and usually a little over 1 inch long and about ½ inch wide. It is similar to Chaparral Honeysuckle in flower color and size, but, unlike that species, the members of its upper pairs of leaves are completely distinct from one another.

Southern Honeysuckle

Poison Oak

Shinyleaf Barberry

Chaparral Flowering Ash

Bladderpod

PLATE 1

Creosote Bush

Squaw Bush

Armed Senna

Mesa Dalea

PLATE 2

Burrobush

White Burrobush

California Beloperone

Bladder Sage

PLATE 3

Desert Sage

Cleveland Sage

Woolly Blue-curls

Desert Willow

PLATE 4

Ocotillo

Prickly Phlox

Tree Tobacco

Western Azalea

PLATE 5

Chaparral Nightshade

Desert Holly

Bush Chinquapin

California Buckwheat

PLATE 6

Interior Live Oak

Chamise

Red Shanks

Wartleaf Ceanothus

PLATE 7

Toyon

Bitterbrush

Laurel Sumac

Gooseberry

PLATE 8

Bush Monkeyflower, Sticky Monkeyflower
(Mimulus)

The Bush Monkeyflowers are a group of closely related species often abundant on brushy slopes throughout the western part of our area. They are sometimes segregated as a separate genus, *Diplacus*. The English name of these plants is derived from the supposed resemblance of the corolla, viewed head on, to a monkey's face. All the bushy species of *Mimulus* flower most abundantly in spring and early summer. *Mimulus puniceus* has red flowers; the others have yellowish ones. The upper stems and calyces of *M. longiflorus* and *M. calycinus* are covered with short hairs; those of *M. aridus* and *M. aurantiacus* are not. Our species will be treated under three headings.

Red Monkeyflower *(Mimulus puniceus)*

Red Monkeyflower is a handsome shrub with leaves mostly about 1½ inches long and ½ inch wide. Its leaves are dark green and smooth above, paler green below. The bright red flowers are about 1½ inches long and borne in leaf-axils. Red Monkeyflower is found on brushy slopes from Ventura County southward, not occurring far from the ocean. This species forms evident hybrids with several of the yellow-flowered species; and, particularly in San Diego County, a single hillside may be covered with plants having flowers of every conceivable intermediate hue between bright red and distinct yellow. Toward the northern part of its range, red-flowered plants are rare, and apparently many of them have resulted from ancient hybridization with one of the yellow-flowered species, usually the common Southern Monkeyflower, *Mimulus longiflorus*. Apparently, however, the beautiful red-flowered form of Southern Monkeyflower that grows in the vicinity of Santa Susanna Pass, Ventura County, is an independent variation that has nothing to do with hybridization involving *M. puniceus*.

Red Monkeyflower

Southern Monkeyflower *(Mimulus longiflorus)*

In this, the commonest Bush Monkeyflower over much of coastal southern California, the young branches and underside of the leaves are covered with short sticky hairs. The yellowish green leaves are usually about 2 inches or more in length and only about ½ inch wide. The flowers are mostly over 2 inches long and clear or pinkish-yellow. Southern Monkeyflower is widespread throughout the coastal part of our area. The closely similar *Mimulus calycinus,* which is lower in habit and has cream-colored corollas, occurs on dry rockslides and crevices in pine forest and desert woodland. Its range includes the San Bernardino, San Jacinto, and Santa Rosa mountains, as well as the

Southern Monkeyflower

Joshua Tree National Monument—all regions where Southern Monkeyflower is absent. It also occurs with that species, but at higher elevations, in the San Gabriel, Santa Ana, and Cuyamaca mountains.

Northern Monkeyflower *(Mimulus aurantiacus)*
The foliage of Northern Monkeyflower lacks the covering of sticky, densely matted hairs found in Southern Monkeyflower, and its leaves and flowers are smaller. It reaches our area in the coastal mountains from Point Conception northward. Another more local

[67]

species also lacks the dense covering of hairs found in Southern Monkeyflower, but its flowers are generally more than 2 inches long. This is *Mimulus aridus,* which is locally common on dry rocky slopes in the southeastern part of San Diego County and also on the desert slopes of the Laguna Mountains.

Whorl-leaf Penstemon *(Penstemon ternatus)*

This attractive shrub with graceful arching branches is often about 6 feet tall. Its toothed leaves are about 1 inch long and half as wide, and arise in groups of 3 at each node. The dull orange-red flowers are about 1 inch long, and are borne in large loose clusters at the ends of the branches. This attractive summer-flowering shrub is one of the few woody plants of our area that have whorled leaves. It is fairly frequent on brushy slopes or in oak woodland from the Santa Ynez and San Rafael mountains of Santa Barbara County southward.

Climbing Penstemon *(Penstemon cordifolius)*

Climbing Penstemon is a straggling shrub which sprawls over other plants on brushy slopes. Its minutely toothed leaves are mostly about 1½ inches long and half as wide, and are truncate or somewhat heart-shaped at the base. The attractive scarlet flowers are about 1½ inches long, and borne in short leafy clusters at the summit of the stem. They are conspicuously 2-lipped and appear in spring. Climbing Penstemon is frequently encountered on brushy slopes below 4,000 feet throughout southern California except for the deserts.

Yellow Penstemon *(Penstemon antirrhinoides)*

A well-branched woody shrub, often about 4 feet tall, Yellow Penstemon is fairly common on chaparral-covered slopes from the southern foothills of the San Bernardino Mountains southward. It reaches the margins of the deserts but does not extend near the

Climbing Penstemon

coast. Its narrow leaves are less than 1 inch long, and its gaping yellow flowers are about ½ inch long. They appear in spring, and are clustered at the tops of the stems. The similar but white-flowered *Penstemon breviflorus* is found in the higher interior Coast Ranges from northern Los Angeles County northward. Rothrock Penstemon, *Penstemon rothrockii,* is a low but distinctly woody yellow-flowered species found locally in the San Jacinto Mountains between 7,000 and 9,000 feet elevation. Its small flowers are not stalked like those of Yellow Penstemon. A number of other nonwoody species of Penstemon are common in southern California.

Bladder Sage *(Salazaria mexicana)*

Bladder Sage is a low desert shrub with stiff branches that become somewhat spiny in age. Its smooth leaves are about ½ inch long and are narrow; when crushed they exude a strong minty odor. The calyx becomes inflated and bladder-like in fruit, and approaches 1 inch in length; the dry sacs resulting from this are a very conspicuous feature of the fruiting plants. The sacs often become tinged with rose in age. The attractive purple flowers are about ½ inch long, and very irregular in form. Bladder Sage is common on rocky slopes through both our deserts.

[69]

Sage *(Salvia)*

Although Zane Gray's *Riders of the Purple Sage* were galloping through stands of what is often called Sagebrush *(Artemisia),* the true sage of the Mediterranean region is a close relative of the shrubs of the Mint Family and will be discussed next. Collectively, the members of this group can be recognized by their two stamens, opposite leaves, and minty odor. It is very interesting, however, to notice the great differences in odor between the various species of Sage. The species of *Salvia* are in general very distinct from one another, but where they come together, particularly in somewhat disturbed situations, hybrids may be formed between them. Such hybrids can usually be recognized very easily by their intermediate characteristics, and they can often be found by searching carefully in areas where two species of Sage occur together. Species of *Salvia* bloom in late spring and early summer.

Greata Sage, our rarest species, is found only in a few canyons along the northeast side of the Salton Sea. It can easily be recognized by its gray-green leaves which bear several long spine-tipped teeth along their margins, and by its pale lavender-blue flowers. In periods of drought the plants lose all their leaves and are difficult to find even in localities where they are known to be present.

Only two closely related species of all the rest have entire leaves: Desert Sage, *S. dorrii,* and Mountain Desert Sage, *S. pachyphylla.* Both have thin, paperlike, usually purplish bracts among their clusters of deep or violet-blue flowers. Their leaves, which are about ½ inch long in Desert Sage and longer on Mountain Desert Sage, are about half as wide and are broadest in their upper half. The leaves in all our other species are broader near the base than in the upper half. Desert Sage occurs above 2,000 feet on rocky slopes in the mountains around the western and southern end of the Mojave Desert. Mountain Desert Sage, distinguished principally by its larger bracts and leaves, is found on rocky, often loose slopes above 5,000 feet in the San Bernardino and San Jacinto mountains.

White Sage *(Salvia apiana)*

One of our most conspicuous shrubs, White Sage is erect and reaches over 6 feet in height. In all other southern California species of sage, the flowers are bunched in dense clusters along the upper part of the stem; but in White Sage, they are loosely grouped in full bunches up to 4 feet long. The leaves are up to 4 inches long and usually about a third as wide, shallowly toothed along the margins, and somewhat whitish, being densely covered with short appressed hairs. The flowers are white and of an elaborate shape, with an elongate lower lip, and sometimes have small lavender dots or streaks. They are mostly about ¾ inch long, and, like those of other species of the group, appear in late spring and early summer. White Sage occurs throughout southern California from Santa Barbara County south, ranging eastward to the margins of the desert.

Sand Sage *(Salvia eremostachya)*

Sand Sage and Mojave Sage can be distinguished from the remaining species by the thin, rounded, paper-like, often purplish bracts subtending the clusters of flowers. In all other following species of *Salvia* the bracts are reduced and inconspicuous. The narrow leaves of Sand Sage are about 1 inch long, and their upper surface is densely covered with bumps. The purplish-blue flowers are about ¾ inch long. Sand Sage is local and scattered along the western edge of the Colorado Desert, but frequent at certain localities along the Palms to Pines Highway.

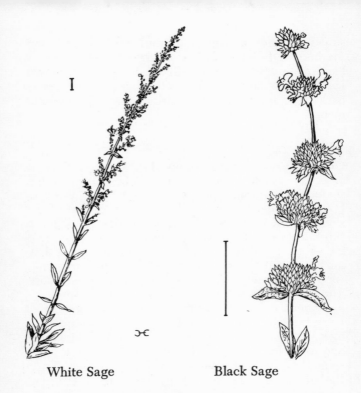

White Sage   Black Sage

Mojave Sage *(Salvia mohavensis)*

In Mojave Sage, a rather rare shrub found in the Joshua Tree National Monument and eastward in the desert mountains of San Bernardino and Riverside counties, a single flower cluster terminates each branch. The pale lavender-blue corollas are nearly hidden by the large rounded whitish bracts. The leaves of Mojave Sage are narrow and mostly about ½ inch long.

Bristle Sage *(Salvia vaseyi)*

In Bristle Sage, aside from White Sage the only white-flowered species of our area, each lobe of the calyx is tipped by a long needle-like bristle. Its leaves are 1 to 2 inches long and ½ to 1 inch wide. The flower clusters are arranged along the upper part of the stems from up to 2 feet. Bristle Sage is found on washes and hillsides from Morongo Valley to the western edge of the Colorado Desert, never far from the desert.

Purple Sage *(Salvia leucophylla)*

The grayish leaves of Purple Sage are distinctive, being abruptly truncate at the base rather than tapering gradually into the leaf stalk as in Cleveland Sage and Black Sage. These leaves are often about 2 to 3 inches long and less than a third as wide, finely wrinkled above and densely covered with short grayish hairs. The purplish flowers are grouped into 3 to 5 clusters, and are about ½ to ⅝ inch long. Purple Sage is found on hillsides near the coast from Pismo Beach, San Luis Obispo County, to Santiago Canyon, Orange County.

Cleveland Sage *(Salvia clevelandii)*

Cleveland Sage is distinguished from Black Sage by its large blue flowers, with corollas about ¾ inch long. They are borne in a single cluster or very few widely spaced clusters. The leaves are about 1 inch long and half as wide. Cleveland Sage is locally common on brushy slopes in western San Diego County, where it forms a spectacularly distinct hybrid with the very different White Sage when the two come together. It is perhaps our most aromatic species.

Black Sage *(Salvia mellifera)*

In the common Black Sage, the corolla is less than ½ inch long. The flowering stems are surmounted by numerous small closely spaced clusters of blue flowers. The leaves of Black Sage are mostly 1 to 2 inches long and about a third as wide. Black Sage is probably our commonest species of *Salvia*, being abundant on grassy hillsides below 2,000 feet eastward to the margins of the desert. In southwestern San Diego

County, from San Miguel Mountain and the region of Jamul southward, it is replaced by the closely similar *S. munzii,* which has leaves mostly about ½ inch long. *Salvia brandegei,* a third member of this immediate group, is found only on Santa Rosa Island; Black Sage itself is found on the other islands. Black Sage is considered an excellent native source of honey by beekeepers.

Common Snowberry *(Symphoricarpos albus)*

Common Snowberry is a trailing shrub, or up to about 4 feet tall, found in somewhat moist shaded situations throughout the western part of southern California away from the deserts. The broad leaves are mostly about 1 inch or less in length and rounded at both ends, with the margins entire. They are dull green above and paler below, being more or less covered with fine hairs. The flowers, which appear in spring, are bell-shaped and white, tinged with pinkish; they range up to ¼ inch long. More conspicuous are the soft pure white berries, up to ½ inch thick, which are borne in clusters and are often very evident in fall and winter, particularly after the leaves of this deciduous shrub have been shed.

Mountain Snowberry *(Symphoricarpos vaccinioides)*

Unlike Common Snowberry, Mountain Snowberry has trumpet-shaped flowers ¼ inch or more in length. They are white, yellowish, or pink, and appear on the shrubs in early summer. The leaves are dull green above and paler below, and are similar to those of the preceding, although generally smaller. They are also less truncate at the base. Mountain Snowberry is fairly common on dry rocky slopes above 5,000 feet in most of the high mountains of southern California.

Woolly Blue-curls *(Trichostema lanatum)*

Plate 4 should serve to distinguish this handsome shrub from any other member of our native flora. The

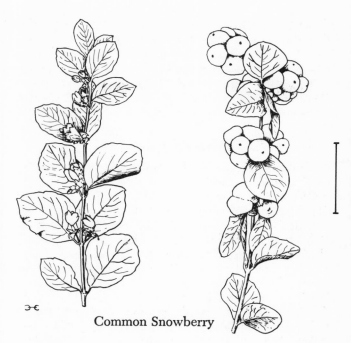

Common Snowberry

plants themselves range up to about 3 feet in height, and are rather densely covered with narrow leaves, dark green above and densely covered with short matted hairs below. They are up to about 2 inches in length, and have infolded margins. The elaborate 2-lipped purple flowers are about 1 inch long, with the fringed lower lip being longer than the tube. These flowers are densely grouped along the upper stem, and the whole cluster is covered with dense purple wool. The stamens stick out of the corollas for a distance of about 1½ inches. Woolly Blue-curls is abundant on brushy or disturbed slopes, as along roadsides, throughout the western part of southern California. Parish Blue-curls, *T. parishii,* is similar, but has smaller flowers much more loosely clustered—borne, in fact, mostly on secondary branches rather than on the main stem axis.

Parish Blue-curls occurs from about 2,000 to 6,000 feet elevation, mostly near the margins of the desert. from near Acton, Los Angeles County, to San Diego County, where on Otay Mountain it comes together with Woolly Blue-curls, which is on the lower slopes. Woolly Blue-curls is rarely found above 2,500 feet elevation.

GROUP VIIIB

Shrubs with petals present and united, at least at the base, and leaves alternate.

Of the genera belonging to this group, four are distinctly spiny: *Fouquieria* (Ocotillo), a tall often leafless desert shrub with red flowers; *Leptodactylon,* which has needle-like leaves and pink or purplish flowers; *Menodora,* in which the flowers have only 2 stamens; and *Lycium,* with 5 stamens and fleshy leaves.

Most of the remaining genera are members of the Heather Family: *Arctostaphylos, Comarostaphylis, Rhododendron, Vaccinium, Xylococcus.* The entire group can be divided, as McMinn has done, into shrubs in which the leaves are toothed along the margins and those in which they are entire (not toothed), except occasionally in vigorous shoots. Among the group in which the leaves have entire margins, *Chilopsis* is a desert shrub with highly irregular 2-lipped flowers 1 to 2 inches long. In *Rhododendron,* a shrub of streamsides in the high mountains, and *Styrax,* a shrub of dry mountainsides, the flowers are white and ½ inch to almost 2 inches long. *Nicotiana* has yellow tubular flowers about 1½ inches long. *Solanum* has flat, saucer-shaped flowers in which the stamens are plainly visible from the side, whereas in *Arctostaphylos* (Manzanitas) and *Xylococcus* the flowers are urn-shaped and enclose the stamens.

In the remaining genera the leaves are toothed along their margins. In *Eriodictyon* the flowers are in large terminal clusters; in *Comarostaphylis, Gaultheria,* and *Vaccinium* they are in small lateral bunches.

Manzanita *(Arctostaphylos)*

One of the most important groups of shrubs in the chaparral areas of southern California are the manzanitas. The English name is derived from the Spanish for "little apple," alluding to the small apple-like fruits. The long Latin name, which can be pronounced easily if divided into syllables, comes from two Greek words meaning "bear berry." The white flowers, often tinged with pink, are like small inverted urns, and appear in winter. The thick leathery leaves are fairly characteristic, as is the often reddish bark which peels off in strips. In the related *Xylococcus* the margins of the leaves are conspicuously folded under, something not found among the true manzanitas. Although it is often rather difficult to distinguish species of Manzanita, their abundance and attractiveness make the effort worthwhile. One of our commonest is Eastwood Man-

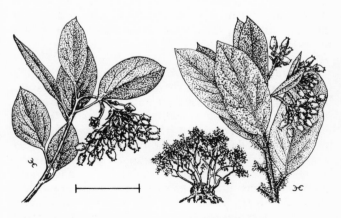

Mexican Manzanita   Eastwood Manzanita

[77]

zanita, *Arctostaphylos glandulosa,* which can be recognized by the large woody burl on its stem just at ground level. None of our other common species has such a burl or woody platform. Eastwood Manzanita sends up new shoots from the base following a fire, whereas the others sprout from seeds. Eastwood Manzanita has smooth reddish bark and leaf-like green bracts among its flowers. Its branchlets are more or less densely covered with sticky spreading hairs. The plants range from 2 to 4 feet in height. This species is common throughout our area on brushy slopes at medium and low elevations, extending east to the margin of the desert.

Perhaps the commonest of the remaining species, at least at low and medium elevations, is Bigberry Manzanita, *A. glauca.* Bigberry Manzanita can be distinguished from all others by its broad grayish-green leaves and large juicy berries, which are about ½ inch thick and have a solid, indivisible stone within. It is a tall erect shrub or often even a small tree up to 25 feet tall. Its branchlets are usually smooth, but a curious hairy form occurs near San Marcos Pass in Santa Barbara County. This species has approximately the same range as the preceding, and forms occasional hybrids with it.

In Lompoc Manzanita, *A. viridissima,* which occurs in northwestern Santa Barbara County, the leaves are heart-shaped at the base and not stalked. Its branchlets are covered with spreading bristly hairs. The remaining three species that will be discussed are found mostly above 3,000 feet in the mountains, often in pine forest. Pinkbract Manzanita, *A. pringlei,* found in the San Bernardino, San Jacinto, and Cuyamaca mountains, has grayish-green leaves and branchlets covered with spreading sticky hairs. Parry Manzanita, *A. parryana,* is similar, but has bright green leaves and may lack the spreading hairs. It ranges from Mount

Pinos to the San Jacinto Mountains. Mexican Manzanita, *A. pungens*, also has bright green leaves, but its branchlets are smooth or finely hairy; its range extends from the San Jacintos to the San Gabriels, and it is also found in the southeastern quarter of San Diego County.

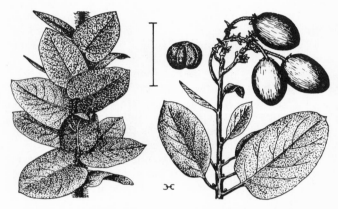

Lompoc Manzanita   Bigberry Manzanita

Desert Willow *(Chilopsis linearis)*

Desert Willow is a tall well-branched shrub of the desert washes, ranging from about 6 feet to over 20 feet in height. Far from being a true willow, it is a member of the largely tropical Bignonia Family. It is deciduous, and the leaves are narrow and willow-like, mostly about 4 to 6 inches long, and nearly smooth. The handsome flowers are in small clusters terminating the branches. There are whitish markings, or often washed with pink or purplish, with darker fine purple markings, highly irregular, and about 1½ inches long. The capsules are also very distinctive, being slender and from 6 to 9 inches long, and containing a number of flat seeds about 1/3 inch long with a tuft of hairs at either end. Desert Willow blooms throughout the warm part of the year and occurs over the Colorado Desert and dry interior valleys to its west, and in the southern Mojave Desert.

Summer Holly *(Comarostaphylis diversifolia)*

A very attractive shrub found locally in ranges near the coast: Santa Ynez Mountains, Santa Monica Mountains, and San Diego County from the vicinity of La Jolla southward. Summer Holly is also found on some of the islands off the coast. In general appearance, it is somewhat similar to the manzanitas, but is easily distinguished from them by the even, sharp toothing along the leaf margins. The urn-like white flowers are borne in small clusters, and appear in spring, while the bright red berries, as the English name implies, appear in summer. Summer Holly is fairly widely cultivated in gardens and could well be used even more widely. Botanically it has about as much in common with the madrone *(Arbutus)* as it does with the manzanitas.

Yerba Santa *(Eriodictyon)*

The various species of yerba santa are a conspicuous feature of the brushy hillsides in many parts of southern California. All bloom in spring. They are generally erect, aromatic evergreen shrubs with rather thick, leathery leaves. The clustered flowers are funnel-shaped and erect, generally about ¼ to ½ inch long, and white or variously washed with purplish. In the first two species of *Eriodictyon* discussed, the leaves are densely covered with short gray hairs on both surfaces, whereas in the remainder of the genus, the leaves are smooth and not hairy, at least on the upper surface. All species of yerba santa flower in spring and early summer. As a group, they occur in disturbed ground or on burns rather than in dense brushland communities. Some of them can actually become pests in planted areas, as they spread by underground runners.

Thickleaf Yerba Santa *(Eriodictyon crassifolium)*

A shrub 4 to 10 feet tall, densely covered with shaggy gray hairs. The leaves are about 2 to 6 inches long and ½ inch to 2 inches wide, and their margins are shallowly toothed. The pale bluish-purple flowers are from ¼ to ⅝ inch long, the corollas open and funnel-shaped. Thickleaf Yerba Santa occurs from western San Diego County north to Matilija Canyon in Ventura County and the San Gabriel Mountains.

Trask Yerba Santa *(Eriodictyon traskiae)*

Trask Yerba Santa is not as tall as the preceding species, but, like it, is completely covered with shaggy gray hairs. Its leaves are 2 to 4½ inches long and ½ inch to 1½ inches wide, with shallowly toothed margins. The corollas are purplish and, unlike those of Thickleaf Yerba Santa, constricted at the summit; they are about ¼ inch long. Trask Yerba Santa occurs on Santa Catalina Island and from the Santa Ynez Mountains of Santa Barbara County northward.

San Diego Yerba Santa *(Eriodictyon lanatum)*

Unlike the preceding species, San Diego Yerba Santa has the leaves smooth and sticky above, but, like them, they are densely gray-hairy below. The leaves are about 1 to 3 inches long and ¼ to ¾ inch wide, shallowly toothed or entire. The purplish or blue corollas are ¼ to ⅝ inch long and not constricted at the summit. San Diego Yerba Santa ranges from the interior of San Diego County to the Santa Rosa Mountains of Riverside County. A somewhat similar species, in which the leaves are smooth but not sticky above, is Santa Barbara Yerba Santa, *Eriodictyon denudatum,* which has lavender corollas of the same size and similar outline. It is found in Santa Barbara and Ventura counties.

Smooth Yerba Santa *(Eriodictyon trichocalyx)*

In Smooth Yerba Santa the leaves have only very fine short hairs below. They are 2 to 5 inches long and

½ inch to 1½ inches wide, being quite willow-like in outline. As they are sticky, they catch a great deal of the summer dust, and the plants often appear ragged and dirty. They are attacked, often severely, by a black fungus which discolors them. The flowers are pale lilac or white, ¼ to ⅜ inch long, and funnel-shaped. Smooth Yerba Santa is common from the Santa Ynez Range south along the desert slopes of the mountains to Riverside County. Lompoc Yerba Santa, *Eriodictyon capitatum,* is somewhat similar, but has entire leaves only about ⅛ inch wide and flowers crowded into head-like terminal clusters. It occurs only on disturbed slopes in the coastal forest in the mountains north of Lompoc, Santa Barbara County.

San Diego Yerba Santa

Ocotillo *(Fouquieria splendens)*

The tall, whip-like, clustered branches of the Ocotillo are a familiar sight throughout the Colorado Desert and into some of the interior valleys to its west. In the warm part of the year after rains the spiny branches are often covered with bright green leaves about 1 inch long. At such times clusters of brilliant red flowers about 1 inch long appear along the terminal segments of the branches. They are tubular and have 5 petals, and are visited by hummingbirds. Ocotillo is absolutely unmistakable. It rarely flowers when less than the height of a man.

Salal *(Gaultheria shallon)*

Salal is an attractive spreading evergreen shrub of the northern woods, with dark green leathery leaves, mostly 2 to 4 inches long and 1 to 1½ inches wide. The leaves are rounded or heart-shaped at the base and finely toothed all around. The urn-shaped white flowers are 1/3 inch long or a little longer, sometimes attractively tinged with pink, and are followed by black fruits ¼ inch or more in diameter. The leaves of Salal are much used by florists, especially for Christmas decorations. It is always found near the coast and reaches its southern limit in the Santa Ynez Mountains west of Santa Barbara. Salal flowers in spring.

Prickly Phlox *(Leptodactylon californicum)*

Prickly Phlox is a densely leafy shrub rarely more than 3 feet tall. Its needlelike leaves are deeply divided and closely bunched all over the stems. In spring and early summer the whole shrub is often covered with attractive pink or rosy flowers more than 1 inch across. They are saucer-shaped, but have a slender tube ½ to ¾ inch long beneath the spreading corolla lobes. Prickly Phlox grows on dry brushy hills from San Luis Obispo County south to the Santa Monica Mountains, and to the San Bernardino, San Jacinto, and Santa Ana mountains farther south.

Desert Thorn *(Lycium)*

As a group, the species of Desert Thorn can be recognized by their spiny branches, their juicy succulent leaves which are usually about 1 inch or less in length, and their tubular flowers. They might be confused with another spiny desert shrub, *Prunus fasciculata,* which, however, has separate petals and dry, peachlike fruits, rather than the rounded berries found in Desert Thorn. The species of Desert Thorn are often very difficult to distinguish, even for a specialist, but it is hoped that the following notes will be of assistance. I shall divide the species of *Lycium* into two groups, the usual procedure. In plants of Group A the lobes of the corolla are at least two-thirds as long as its tubular portion; in Group B the lobes are less than two thirds as long as the tube. In plants of Group A the corolla lobes are at least 1/16 inch long, whereas in those of Group B they are this length or shorter.

LYCIUM, GROUP A

Corolla lobes two-thirds as long as the tube or longer.

Rabbit Thorn *(Lycium pallidum)*

Rabbit Thorn is a spiny shrub 2 to 6 feet tall. The leaves have a thin waxy covering which makes them appear grayish-green. The flowers are ½ to ¾ inch long and trumpet-shaped, 3/16 inch or more in diameter at the summit, and white or lavender in color. In other species of Group A, the corollas are more or less tubular and of a smaller diameter at the summit. The bluish berry is globose. Rabbit Thorn flowers in early spring, and is found from the central and western Mojave Desert to Death Valley.

Cooper Desert Thorn *(Lycium cooperi)*

Cooper Desert Thorn is about the same height as the preceding, but lacks the waxy coating on the leaves and has smaller tubular flowers. Its most distinctive characteristic is its fruit, which is sharply constricted all around just below the summit. Cooper Desert Thorn flowers in early spring, and is found throughout both deserts.

Desert Thorn *(Lycium brevipes)*

Desert Thorn is similar to the two preceding species, but its pink or violet flowers are funnel-shaped, neither tubular nor trumpet-shaped, and are less than ⅜ inch long. Its leaves lack a waxy covering also, and its berry is globose and red. Desert Thorn flowers in late winter and occurs on the Colorado Desert.

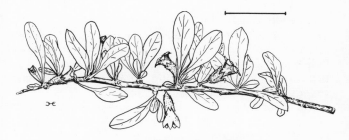

Cooper Desert Thorn

LYCIUM, GROUP B

Corolla lobes less than two-thirds as long as the tube.

Coast Desert Thorn *(Lycium californicum)*

Distinctive because of its coastal habitat, Coast Desert Thorn is often matted and less than 2 feet tall. It ranges along the coast from the Santa Monica Mountains southward, and flowers in late winter. Its bright red berries have only 2 seeds, which is unique among our species, most of which have many seeds in each berry. After its small narrow leaves fall, the branches are marked by corky bumps. The white flowers of this species are less than ¼ inch long, and have only 4 corolla lobes instead of the 5 commonly found in other species.

[85]

Anderson Desert Thorn *(Lycium andersonii)*

Aside from the preceding, Anderson Desert Thorn is the only species of its group with leaves regularly less than ⅛ inch wide. Even though individual leaves on the other species may not be this wide, a number on any given bush are. Anderson Desert Thorn lacks the corky bumps found on the branches of Coast Desert Thorn. Its pale lavender flowers appear in winter, and are ¼ to ⅜ inch long. This species ranges from coastal southern San Diego County to the Colorado and Mojave deserts.

Fremont Desert Thorn *(Lycium fremontii)*

This shrub of the Colorado Desert can be distinguished by the dense covering of hairs all over its leaves. Other species of its group may have a few hairs on the young leaves, but nothing approaching the dense covering of Fremont Desert Thorn. The flowers are white to somewhat lavender, 5/16 to ⅜ inches long, and appear in late winter. The berries are red. Fremont Desert Thorn occurs on the Colorado Desert.

Torrey Desert Thorn *(Lycium torreyi)*

In Torrey Desert Thorn the mature leaves are smooth or at most very slightly hairy. The flowers are similar to those of the preceding, but the calyx tube is 3/16 to 5/16 inch long, whereas in the preceding it is shorter. In *Lycium brevipes,* which might be placed in Group B depending on the individual examined, the flowers are less than ⅜ inch long; in the present species, longer. Torrey Desert Thorn ranges from the Colorado Desert to the vicinity of Needles, and blooms in spring.

Spiny Menodora *(Menodora spinescens)*

An intricately branched spiny desert shrub less than 3 feet tall with leaves less than ½ inch long. The white flowers, ⅛ to ¼ inch long have only 2 stamens, a characteristic that allows *Menodora* to be separated from all other similar plants. Spiny Menodora blooms in spring and occurs in the eastern Mojave Desert, where it is not infrequent.

Tree Tobacco *(Nicotiana glauca)*

This native of temperate South America has become one of the most familiar shrubs throughout the western part of our area. It ranges from 6 to 15 feet in height, and flourishes along streams, but is also found generally in waste or disturbed ground. The gray-green undivided leaves are 2½ to 5 inches wide, on a stalk 1 to 2 inches long. The yellow flowers, which are borne in loose terminal bunches, are narrowly tubular and about 1½ inches long. The dry fruits contain numerous very small seeds.

Western Azalea *(Rhododendron occidentale)*

Our native Western Azalea is a very beautiful shrub of streamsides in the San Jacinto Mountains and higher ranges of San Diego County. Its flowers are white, often with a pinkish tinge, and with a yellow blotch on the upper part, and are 1½ to 2 inches long. They are somewhat irregular, and the 5 stamens are exserted and plainly visible. The leaves are soft, mostly 2 to 3 inches long and ½ to 1 inch wide. Western Azalea flowers in spring and early summer.

Chaparral Nightshade *(Solanum xanti)*

Although several species of nightshade are recognized from southern California, the distinctions between them are so imperfectly understood that we shall consider them collectively under this heading. All are low shrubs, rarely more than 3 feet tall, and only slightly woody at the base, and most of them exude a rank odor when crushed. The leaves are mostly ½ inch to 2½ inches long and somewhat less than half as wide, with entire or sometimes slightly irregular markings. In most of our representatives they are thickly clad

[87]

with soft hairs. The saucer-shaped flowers are mostly violet or lavender, and the 5 yellow stamens are erect and clustered around the pistil in the center. The berries are mostly green or purplish, up to ½ inch or more thick, and, like the leaves, poisonous. These relatives of the cultivated potato and tomato grow throughout southern California except in the deserts and high mountains.

Snowdrop Bush *(Styrax officinalis)*

Snowdrop Bush is a tall attractive shrub, mostly 4 to 10 feet tall, which flowers in spring and grows on dry bushy slopes at scattered localities from San Luis Obispo County south to San Diego County. It is deciduous, and the softly hairy leaves are mostly 1½ to 3 inches long and nearly as wide. The white flowers are borne in small clusters, and are about 1 inch long, with the white petals, joined only near the base, commonly 6 in number but ranging from 4 to 8. Snowdrop Bush is apparently absent from the Santa Monica and San Gabriel mountains.

California Huckleberry *(Vaccinium ovatum)*

A tall erect shrub with leathery shining evergreen leaves ½ inch to 1¼ inches long and ¼ to ½ inch wide. California Huckleberry has the urn-shaped flowers characteristic of many members of the Heather Family. They are pinkish, about ¼ inch long, and borne in small clusters beneath the more or less horizontal, densely leafy branches. This shrub flowers in spring, and in early summer bears its delicious bluish-black berries which contain numerous tiny seeds. California Huckleberry occurs in the outer Coast Ranges from the vicinity of Santa Barbara northward, and also locally near Escondido, San Diego County.

Mission Manzanita *(Xylococcus bicolor)*

Xylococcus is similar to the true manzanitas *(Arctostaphylos)* in general appearance, but can easily be

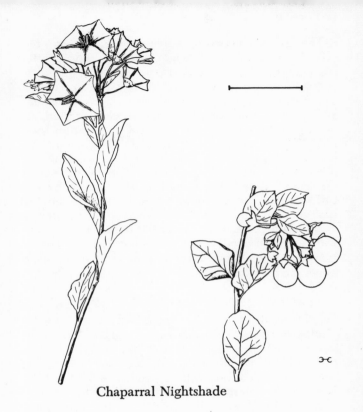

Chaparral Nightshade

distinguished by the margins of its leaves, which are
folded under, and by its dry, hard fruits. Like the
manzanitas, it blooms in late winter and has urn-
shaped white flowers. Mission Manzanita occurs only
on Santa Catalina Island, in the Verdugo Hills, in
western Riverside County, and throughout western
San Diego County, where it is abundant in the coastal
chaparral.

GROUP IX

Shrubs with petals present or absent but not joined
into a tube; the leaves opposite, simple.

This group comprises twelve genera. *Larrea,* creosote bush, would be sought here if the 2-parted compound leaves in opposite pairs were thought to be deeply 2-lobed simple leaves. It is treated on page 28. Three genera in this group, *Adolphia, Coleogyne,* and *Shepherdia,* have spiny branches. In *Galium* there are more than two leaves at each point of insertion. *Acer* (Maple), one kind of *Ceanothus,* and *Euonomys* have teeth along the margins of the leaves. *Garrya* has the flowers hanging in long tassels. The leaves of the species of *Ceanothus* with opposite leaves have a corky brown raised bump on each side of the leaf stalk at its base. All species of *Ceanothus* are treated together on pages 107-111. The leaf stalks of *Cneoridium* and *Simmondsia* are very short (less than 1/16 inch long) or absent, whereas those of *Cornus,* dogwood, are over ¼ inch long.

Dwarf Maple, Mountain Maple (*Acer glabrum*)

This large shrub is easily recognized by its typically maple-like leaves with their broad, pointed lobes. The leaves are 1 to 3 inches across and completely lack hairs. The fruits are clustered, and each consists of two halves, each half with a wing ½ to 1 inch long. This species is rare, occurring on moist slopes at middle and high elevations in the San Bernardino and San Jacinto mountains.

California Adolphia (*Adolphia californica*)

A very spiny, intricately branched shrub with small tender leaves about ¼ inch long which readily fall from the branches. The flowers are tiny, greenish white, and inconspicuous. California Adolphia grows on dry brushy slopes in western San Diego County within about 20 miles of the ocean.

Bushrue (*Cneoridium dumosum*)

Strong-scented low shrub with flexible branches and narrow crowded leaves. The leaves are less than 1 inch long. If held up to the light, they can be seen to be dotted with small glands all over the lower surface. The small white flowers, about ¼ inch long, appear in late winter. Bushrue inhabits brushy slopes near the coast in western Orange and San Diego counties.

Blackbrush (*Coleogyne ramosissima*)
Blackbrush is a low spiny shrub, rarely as tall as 6 feet, which has ashy-gray bark that darkens in age. The narrow leaves are less than ½ inch long and fall early. The flowers lack petals, but have 4 yellowish sepals, cupped around the stamens and turning brown in age, and numerous stamens. They appear in spring. Blackbrush occurs in mountain slopes over the Mojave Desert and also on the western margin of the Colorado Desert, and is often locally very abundant at slightly higher elevations than Creosote Bush.

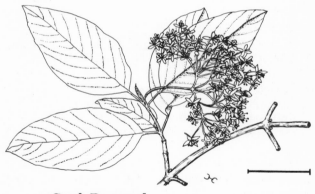

Creek Dogwood

Dogwood (*Cornus*)
Two kinds of shrubby dogwoods grow along the sides of streams and lakes in the western part of southern California. They both have rounded clusters of small

white flowers in late spring and early summer, followed by white or blue berries later in the year. The commoner Creek Dogwood, *Cornus occidentalis,* has leaves 1 to 2 inches long with only 3 or 4 veins on each side of the midrib; the less frequent Smooth Dogwood, *C. glabrata,* has leaves mostly 2 to 4 inches long, with 4 to 7 veins on each side of the midrib.

Western Burning Bush *(Euonomys occidentalis)*

A large shrub of mountain streamsides with soft leaves, the margins of which are finely and minutely toothed all around. The leaves, mostly about 2 inches long and about 1 inch wide, are borne on stalks about ¼ to ½·inch long. The strange purplish-brown flowers, about ⅜ inch across, are borne in late spring in small clusters arising from among the leaves. The large seeds are surrounded by a red fleshy outgrowth. This species is uncommon at about 5,000 feet elevation in the San Jacinto, Palomar, and Cuyamaca mountains.

Bedstraw *(Galium)*

With their whorled leaves and tiny white or greenish flowers, the two semi-shrubby kinds of Bedstraw in southern California can hardly be confused with anything else. The erect *Galium angustifolium,* Chaparral Bedstraw, occurs through the brushy slopes of the western part of our area, and the sprawling *G. stellatum* is found in the desert mountains.

Silktassel Bush *(Garrya)*

Two closely similar species of *Garrya* occur in southern California, which, although they are modally distinct and occupy different ranges, are often very difficult to tell apart where they approach one another. The staminate and pistillate flowers of *Garrya* occur on different bushes, and it is only the latter that bear the chains of berries. *Garrya* species flower in winter. Pale Silktassel, *G. flavescens,* and Veatch Silk-

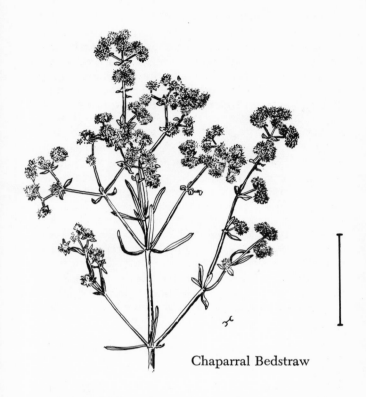

Chaparral Bedstraw

tassel, *G. veatchii,* are for all practical purposes indistinguishable. The lower surfaces of the leaves of the latter are usually covered with a woolly mat of curled or wavy hairs, whereas in the former these hairs are nearly straight and not so dense. Pale Silktassel occurs on dry brushy slopes in the mountains from San Diego County north through the Coast Ranges; Veatch Silktassel occurs throughout the same area. Both species are found above 2,400 feet. From Los Angeles to Santa Barbara County *G. veatchii* often has somewhat undulate leaves like those of the more northern *G. elliptica.*

Silver Buffaloberry *(Shepherdia argentea)*

Erect spiny shrub with shaggy bark. Leaves densely covered with star-shaped hairs on both surfaces, about 1 inch long and about ⅜ inch wide. Flowers small, lacking petals, inconspicuous, the staminate and pistillate flowers on separate plants. Fruit red, berry-like, edible. Silver Buffaloberry occurs on rocky streambanks in the Mount Pinos region and also near Victorville.

Jojoba, Goatnut *(Simmondsia chinensis)*

Tall shrub with thick leathery yellowish-green leaves 1 to 2 inches long and about ¾ inch wide. Flowers greenish, lacking petals, the staminate and pistillate flowers on separate plants. Fruit a smooth, mostly 2-seeded capsule about ¾ inch long. The seeds yield a liquid wax with desirable commercial properties, but it has not yet proved commercially feasible to cultivate this plant. Goatnut is a native of rocky hills along the desert margin in Riverside, Imperial, and San Diego counties, and in southwestern San Diego County near the coast.

Group X

Shrubs which lack petals, thus having only a single whorl of flower parts, the calyx, outside the stamens. The sepals may be brightly colored. Leaves alternate.

Several members of this group are outstandingly distinct: Fremontia, with saucer-shaped yellow flowers at least 1 inch in diameter; *Quercus* (oaks), bearing acorns; *Castanopsis* (chinquapin), with large nuts borne in large spiny burs at least 2 inches across; *Ricinus* (Castor Bean), with maple-like leaves 4 to 10 inches across; and *Cercocarpus* (mountain mahogany), in which the seeds have white feathery tails at least 2 inches long. Others, such as the willow *(Salix)*, are so familiar as to need almost no mention. *Myrica* (wax myrtle) is a tall evergreen shrub with dark green leaves that grows in wet places near the coast from the Santa Monica Mountains north. *Eriogonum* has tiny bisexual flowers gathered into rather showy white clusters. *Rhamnus* is an evergreen shrub with leathery green leaves. The remaining genera are members of the Goosefoot Family and have inconspicuous greenish flowers and more or less grayish foliage. *Allenrolfea* and *Suaeda* are fleshy and somewhat succulent; in *Eurotia* the fruits are covered with long conspicuous silk hairs; in *Grayia* the fruits are enclosed in sacs formed by the fusion of the bracts; *Atriplex* has comparable bracts, but they are not fused.

Bush Pickleweed *(Allenrolfea occidentalis)*

With its jointed fleshy branches, tiny scale-like leaves, and inconspicuous flowers, this shrub of alkaline places on the desert cannot be mistaken for anything else. It is mostly 2 to 4 feet tall.

Saltbush *(Atriplex)*

Our species of saltbush are more or less densely covered with short, mealy white hairs. The inconspicuous flowers are borne in small clusters, and are either staminate or pistillate. Desert Holly, *Atriplex hymenelytra,* which occurs widely on both deserts, has sharply angled leathery leaves as its English name suggests, but, unlike true holly, they are scurfy and dull gray. Parry Saltbush, *A. parryi,* of the Mojave Desert, is the only one of our species that has the leaves heart-shaped at the base. They are less than ½ inch long. In two of the remaining species the leaves are indistinctly stalked, narrowed gradually to the main stem. Both Wingscale, *A. canascens,* and Allscale, *A. polycarpa,* occur throughout the deserts, and the former reaches the coast south of Los Angeles. The fruiting bracts in *A. canescens* have four conspicuous wings, whereas in *A. polycarpa,* which has somewhat broader leaves, the fruiting bracts are merely fringed. Among the species with distinctly stalked leaves, Spinescale, *A. spinifera,* is a very spiny shrub of the western Mojave Desert that normally has the leaves shaped somewhat like a spearhead, whereas Shadscale, *A. confertiflora,* which is found in the same area, has rounded, entire leaves and is usually less spiny. Neither of these species often exceeds 4 feet in height, whereas Lenscale, *A. lentiformis,* which grows throughout southern California, is normally 3 to 10 feet tall. Its leaves are usually broad at the base and sometimes spearhead-like. Lenscale is often planted as a hedge or windbreak along the coast.

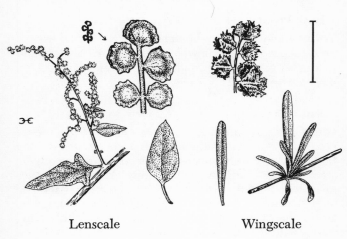

Lenscale   Wingscale

Bush Chinquapin *(Castanopsis sempervirens)*
This evergreen shrub often forms dense colonies on rocky slopes above 6,500 feet in the San Jacinto, San Bernardino, and San Gabriel mountains. Its oblong entire leaves are usually 2 to 3 inches long, green above and more or less golden or pale below. Its large spiny burs are unique among our shrubs.

[96]

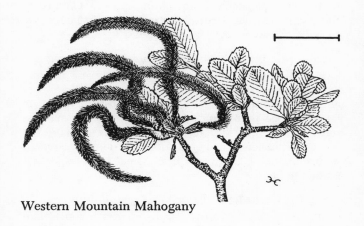

Western Mountain Mahogany

Mountain Mahogany *(Cercocarpus)*

Broadly speaking, there are two very different species of mountain mahogany in southern California. Desert Mountain Mahogany, *Cercocarpus ledifolius,* is a fall shrub with narrow leathery leaves ½ to 1 inch long, the margins entire and rolled under. The tails on the fruits are 2 to 3 inches long. It is common on dry rocky slopes above 4,500 feet in the mountains from Mount Pinos to the San Bernardinos. Western Mountain Mahogany, *C. montanus,* is common everywhere below 6,000 feet, except for the deserts. Its leaves are usually about 1 inch long and half as wide, toothed all around; the tails on its fruits are similar to those of the preceding, with which it cannot be confused.

Ashyleaf Buckwheat *(Eriogonum cinereum)*

Like the other two common shrubby buckwheats, this one has rounded clusters of small white flowers borne on the upper, nearly leafless branches. In Ashyleaf Buckwheat, the margins of the leaves are not folded under, and the leaves themselves are ½ to 1 inch long, ⅜ to ¾ inch wide, and white-hairy below. It is restricted to the coastal bluffs from Santa Barbara County to the vicinity of San Pedro, where it was first collected by a British expedition on October 13, 1839.

California Buckwheat *(Eriogonum fasciculatum)*

Our most abundant shrubby buckwheat, ranging throughout southern California except in the desert and high mountains. The leaves are mostly ½ to ¾ inch long and less than ¼ inch wide, and have the margins folded under. Various forms of this species have the leaves either smooth or more or less densely covered with short hairs. California Buckwheat is an important plant for bees.

Seacliff Buckwheat *(Eriogonum parvifolium)*

Common along the entire coast, but never more than a short distance from the ocean, Seacliff Buckwheat has very distinctive leaves. They are dark green above and densely covered with short white hairs below, somewhat triangular in shape, with a conspicuously infolded margin. Seacliff Buckwheat tends to be more sprawling than our other two species. All three flower throughout the warm part of the year.

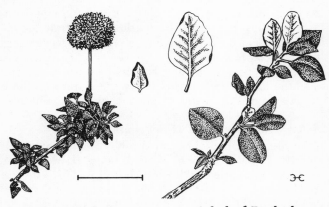

Seacliff Buckwheat   Ashyleaf Buckwheat

Winter Fat *(Eurotia lanata)*

Winter Fat is a low shrub of the western Mojave Desert which might attract attention because of the white tufts on its upper branches formed by the long hairs covering the bracts that enclose its fruit. The leaves are narrow and about 1 inch long.

Fremontia *(Fremontodendron californicum)*

Fremontia, one of our best-known native shrubs, has rapidly gained popularity as an ornamental because of its large yellow flowers 1 to 2½ inches across, lobed dark green evergreen leaves, and fast rate of growth. Under cultivation it flowers throughout the warm part of the year, but in nature it is at its best in late spring and early summer. The thick leaves are densely covered with short star-shaped hairs, and the dry brown fruits are also covered with bristles. Although the 5 sepals are very petal-like, true petals are absent. Fremontia occurs in scattered localities at middle elevations in the mountains throughout our area, but does not range the desert.

Grayia *(Grayia spinosa)*

Grayia is similar in aspect to several species of *Atriplex,* such as *A. polycarpa,* and has somewhat fleshy leaves. The most dependable distinction is found in the bracts, which are completely fused to form open sacs around the fruit. The sacs are ¼ to ½ inch long and usually reddish at maturity. Grayia is found on the Mojave Desert.

California Wax Myrtle *(Myrica californica)*

California Wax Myrtle is a rather uncommon shrub found in the bottoms of moist gullies near the coast from the vicinity of Santa Monica northward. Its tough leaves are dark glossy green, evenly and sharply toothed all around the margins, and mostly 2 to 4 inches long and ½ to ¾ inch wide. The inconspicuous greenish flowers are either staminate or pistillate and borne in small clusters.

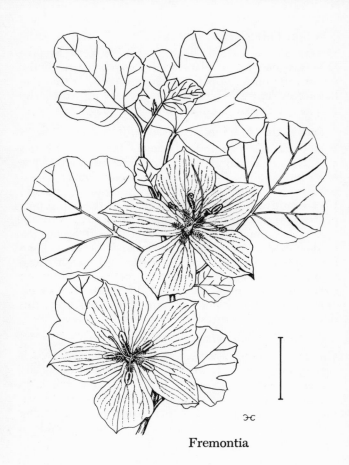

Fremontia

Oak *(Quercus)*

Oaks are familiar to all residents of the North Temperate Zone. They are the only woody plants in southern California that bear acorns, which are nuts with a cup surrounding the point of attachment to the main stem. These acorns come from the pistillate flowers, which are borne in small groups, but the staminate flowers, which are minute and greenish, hang down in slender tassels early in spring. Some of our oaks which are more commonly trees occasionally become shrubby, as California Black Oak, *Quercus kelloggii,* which has deeply lobed leaves, the lobes surmounted by sharp bristles. Other shrubby southern California species have entire or shallowly lobed leaves.

California Scrub Oak *(Quercus dumosa)*

Densely branched evergreen shrub of chaparral-covered slopes, the leaves mostly ½ to 1 inch long and ¼ to ¾ inch wide, gray-green, paler, and often covered with short hairs. The leaves of California Scrub Oak are variable, but mostly are ringed with short irregular spines all around. The acorns mature in a single year. California Scrub Oak is common in the chaparral throughout southern California and local in the desert mountains. The similar Leather Oak, *Quercus durata,* has smaller leaves which are covered with short hairs above and have conspicuous unfolded margins. It is rare in the San Gabriel Mountains and common locally in Santa Barbara County, as on Figueroa Mountain, where it grows on slopes derived from serpentine rock.

Palmer Oak *(Quercus palmeri)*

Tall evergreen shrub of the chaparral from the San Jacinto Mountains south through San Diego County. The leaves are hard and holly-like, ½ inch to 1½ inches long, rarely lacking marginal spines. The acorns mature in their second year, which means that two size classes can be distinguished on the trees, and the scales of their cups are densely and finely hairy. The dull green leaves are lead-colored below or densely short-hairy when young. The similar Gold Cup Oak, *Quercus chrysolepis,* which is normally a tree, occasionally becomes shrublike.

[101]

Interior Live Oak *(Quercus wislizenii)*

Another evergreen shrub of chaparral-covered slopes, Interior Live Oak can be distinguished by its smooth bright yellowish green leaves, which are ¾ inch to 2 inches long and ½ to 1 inch wide, and entire or irregularly spiny around the margins. The acorns require two years to mature. Interior Live Oak occurs as a shrub throughout the western part of our area at middle elevations, and as a tree from Ventura County northward. Coast Live Oak, *Quercus agrifolia,* which rarely has shrubby forms, is similar, but its acorns mature in a single year and the leaves are convex with minute tufts of hair in the axils of the veins beneath.

Castor Bean *(Ricinus communis)*

The Castor Bean is familiar, by sight at least, to every resident of southern California, being frequently planted in gardens and at dooryards and often escaped and established. Its large maple-like leaves are 4 to 10 inches across and their stalk is inserted near the center. The leaves and young branches are juicy and the plants are very fast growing. The flowers are clustered and very inconspicuous; the spiny fruits are nearly 1 inch long. Each contains 3 large shiny mottled brown seeds which are poisonous and unfortunately are occasionally eaten by children. These seeds when pressed yield castor oil. Undoubtedly Castor Bean has been in California since earliest Spanish times; it is found in all warm parts of the globe, but is probably native to Africa.

Redberry *(Rhamnus crocea)*

Sprawling or erect shrub with rounded hard leathery leaves mostly less than 1 inch long. In various races the leaves may be entire or toothed. The flowers are tiny, in small clusters, greenish, with 5 sepals and stamens. The hard red berries are less than ¼ inch across and bright red. Some of the races of this species might be mistaken for oaks if one saw only their leaves, but the flowers and fruits are entirely different. Redberry occurs on brush-covered slopes of low and middle elevations throughout the western part of our area.

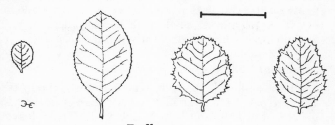

Redberry

Rhamnus crocea   *R. c. pirifolia*   *R. c. pilosa*   *R. c. ilicifolia*

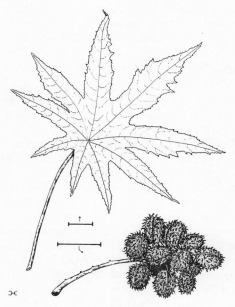

Castor Bean

[103]

Willow *(Salix)*

The narrow leaves and stream or lakeside habitat of willows make them familiar to most people. In willows, the staminate and pistillate flowers occur in dense, narrow clusters on separate plants. When the small fruits on the pistillate plants break open, they release quantities of minute seeds with copious tufts of hair at their base which float lightly through the air.

Sandbar Willow *(Salix hindsiana)*

Sandbar Willow, which is common throughout the western part of our area, has very narrow leaves 1½ to 3½ inches long and ⅛ to ⅜ inch wide. These leaves are more or less densely covered with short silky hairs, and the entire plant has a grayish appearance. Similar is Narrowleaf Willow, *Salix exigua,* which occurs in wet places in or about the deserts. In Narrowleaf Willow the small fruits are smooth, whereas in Sandbar Willow they are covered with short hairs.

Arroyo Willow *(Salix lasiolepis)*

Although sometimes a small tree, this common willow is very often shrubby in habit. Its leaves are 2 to 5 inches long and ⅜ to 1 inch wide, and characteristically broader in their upper half. They are smooth or nearly so, and green in color. Arroyo Willow is found in wet places throughout below 6,500 feet, but not on the deserts. A similar species, Scouler Willow, *Salix scouleriana,* is found in the San Gabriel, San Bernardino, and San Jacinto mountains above 6,500 feet. Its leaves are 1½ to 3 inches long and ¾ inch to 1½ inches wide, broader than those of Arroyo Willow but, like that species, widest in the upper half and smooth. Its fruits are densely hairy; those of Arroyo Willow, smooth. In both species the flower clusters appear on the bare branches before the leaves begin to grow in spring; in three common southern California shrubby willows, the flowers appear after the leaves.

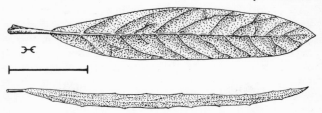

Sandbar Willow

Yellow Willow *(Salix lutea)*

Yellow Willow occurs in moist places in the San Bernardino and San Jacinto mountains above 6,000 feet. It can be recognized by its smooth tapering leaves, which are broadest near the base.

California Sea Blight *(Suaeda californica)*

A low spreading shrub of the coastal marshes, with dense fleshy narrow leaves about 1 inch long. The flowers are green and inconspicuous. Other species of *Suaeda* occur locally around alkaline sinks in the interior.

GROUP XI

Shrubs with both petals and sepals, these differentiated from one another in color, and thus two distinct whorls of flower parts outside the stamens.

Some of the more distinctive members of this group are: *Krameria,* having the petals very unequal to one another in size and hence irregular flowers; *Malacothamnus,* having the stamens fused into a central column; *Petalonyx,* a desert shrub with small white flowers in which the stamens emerge between the petals to a position outside them; *Thamnosma,* a low, highly aromatic purple-flowered nearly leafless desert shrub; *Dendromecon,* which has saucer-shaped bright yellow flowers about 1 inch across; the white-flowered

Lepidium and the red-flowered *Ribes speciosum* have 4 sepals and 4 petals; the remaining genera, 5. *Rhamnus* has tiny inconspicuous greenish flowers; *Adenostoma,* needle-like leaves; and some species of *Ribes,* prickles. In *Ceanothus,* the petals are scoop-shaped, and the main color of the clusters of flowers comes from the blue or white calyx. *Holodiscus, Amelanchier, Prunus,* and *Rhus* have leaves that are entire or merely evenly toothed around the margins; in the other genera the leaves are deeply lobed. In *Rhus* the stamens are 5, and in the other two genera 10 or more. Among the genera with lobed leaves, *Ribes* has 5 stamens, *Heteromeles* 10, and the others 18 or more. *Purshia* has yellow petals and small leaves; *Rubus* has white petals and broad maple-like leaves.

Chamise *(Adenostoma fasciculatum)*

One of the most abundant and distinctive shrubs in the chaparral of southern California, Chamise is likewise one of the most inflammable and contributes largely to the start of brush fires. Often 4 to 8 feet tall, Chamise has needle-like leaves, usually about ¼ inch long, clustered on very short lateral shoots along the main branches. The terminal clusters of small white flowers appear in spring. Chamise is found everywhere on brushy slopes in the western part of our area.

Red Shanks *(Adenostoma sparsifolium)*

Red Shanks is a tall shrub, generally more than 6 feet in height, with slender, needle-like leaves about ½ inch long. The reddish bark of the mature stems characteristically peels off in long strips. The leaves are not clustered like those of Chamise, but are usually borne singly, and the white flowers are borne in looser clusters than in Chamise. Red Shanks flowers in midsummer. It forms colonies in scattered localities throughout the range of Chamise, but above 2,000 feet, and is much less common. In general, its range lies more toward the interior.

Service Berry *(Amelanchier alnifolia)*

An erect deciduous shrub of fairly damp slopes above 4,000 feet in the mountains, Service Berry has rounded leaves about 1 inch long that are evenly toothed around the margins. The white flowers are almost 1 inch in diameter. The purplish fruits are about ¼ inch long, and have the point of insertion of the flower parts at their top, as in an apple, rather than around their base, as in a plum or apricot.

Ceanothus *(Ceanothus)*

There are more kinds of ceanothus native in southern California than of any other group of shrubs. Approximately twenty species of these evergreen shrubs lend beauty to our brushy hillsides with their showy clusters of blue or white flowers. The more common of these will be described below. Since many of the species have a relatively limited distribution, range may often be used to good account in identifying them. Hybrids are known within each of the two groups treated below, but are exceedingly rare between these groups. When two or more different kinds of ceanothus grow together, it is very interesting to search for natural hybrids. In Group B the base of the leaf stalk has no swollen dark corky bumps flanking it, and the leaves are alternate; in Group A such corky bumps are present, and the leaves are opposite or alternate.

CEANOTHUS, GROUP A

Three species of this group have opposite leaves. Hoaryleaf Ceanothus, *Ceanothus crassifolius,* is an erect shrub 3 to 12 feet high, distinguished from other members of the group by its thick leaves, which are densely covered with short white hairs below and usually have sharp coarse teeth on their margins. The leaves are about ½ inch to 1¼ inches long, ⅜ to ¾ inch wide, and the margin is usually infolded all around.

The white flowers appear in late winter. The species is found throughout the mountains of the western part of our area, mostly below 3,500 feet. Buckbrush, *C. cuneatus,* occupies the same area, but is replaced in the desert and desert-border ranges by the scarcely distinct *C. greggii.* Buckbrush does not come near the immediate coast. Its leaves are neither hairy nor toothed, but they are about the same size as those of *C. crassifolius.* The white flowers of Buckbrush appear in early spring.

Two other species of Group A have alternate leaves. Bigpod Ceanothus, *C. megacarpus,* is very common near the coast from near Santa Barbara to the Santa Ana Mountains. Its leaves are about 1 inch long and have entire margins. Its white flowers appear mostly in late winter. Wartystem Ceanothus, *C. verrucosus,* is similar, but has leaves only about ½ inch long, which sometimes have rigid teeth along their margins. It is restricted to western San Diego County south of Encinitas, and thus outside the area of *C. megacarpus.*

CEANOTHUS, GROUP B

Leaves alternate, without swollen dark corky bumps at each side of the base of the stalk.

Within this group, some of the species have 3 distinct main veins from the base of each leaf; 2 of these are distinctly spiny. Snow Bush, *Ceanothus cordulatus,* is a very spiny spreading shrub usually less than 3 feet tall, often abundant on rocky slopes and flats in pine forest above 6,000 feet, from the San Jacinto Mountains north. The entire leaves are mostly ½ to 1 inch long and ¼ to ½ inch wide, and grayish green in color. The white flowers appear from May to July. A second spiny species, Chaparral Whitethorn, *C. leucodermis,* is an erect shrub often approaching 10 feet in height, with leaves similar to those of Snow Bush in shape and size. The pale blue flowers appear in spring. Chaparral

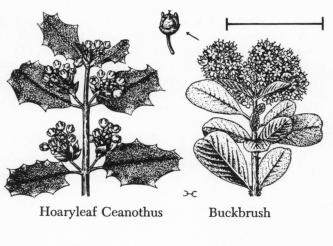

Hoaryleaf Ceanothus   Buckbrush

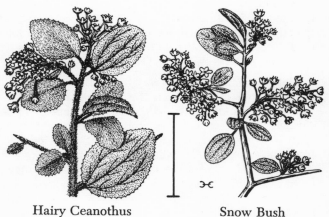

Hairy Ceanothus   Snow Bush

Whitethorn is frequent in chaparral at moderate elevations throughout the western part of southern California.

Nonspiny species with 3 distinct veins on each leaf
include the white-flowered, entire-leaved Deer Brush,

[109]

C. integerrimus, which is a handsome shrub often more than 6 feet tall with large clusters of flowers. Its light green leaves are usually about 1½ inches long. Deer Brush usually occurs in pine forest above 4,000 feet from the San Jacinto and Santa Ana Mountains northward. Other species with distinctly 3-veined leaves have blue flowers that appear in late winter and early spring. Hairy Ceanothus, *C. oliganthus* (including *C. sorediatus* and *C. tomentosus*) is the most common of these, occurring on brushy slopes at low or medium elevations away from the desert. Its leaves are about 1 inch long and ½ inch wide, toothed all around the margins, paler and somewhat hairy on the lower surface. This species has rigid branches that sometimes become rather spiny when the leaves have dropped off, whereas San Diego Ceanothus, *C. cyaneus,* a beautiful deep-blue-flowered species restricted to a limited area about 20 miles northeast of San Diego, and Blue Blossom, *C. thyrsiflorus,* found along the coast of northern Santa Barbara County, have flexible branches that never become spiny, and somewhat larger leaves.

The remaining species of Group B have a single prominent main vein from the base of each leaf. In Wartleaf Ceanothus, *C. papillosus,* the entire upper surface of the leaves is roughly and evenly covered with bumps. These leaves are about 1 inch long and less than ½ inch wide. The dark blue flowers appear in spring. Wartleaf Ceanothus is local in the mountains of Santa Barbara and Ventura counties, and in the Santa Ana Mountains, below 3,000 feet. A similar species with leaves about half as large and lacking the warty appearance is Wavyleaf Ceanothus, *C. foliosus,* which occurs on dry ridges in the Cuyamaca Mountains of San Diego County. In both species the margins of the leaves are toothed.

The three remaining species of ceanothus have entire leaves. The most common of the three is Red-

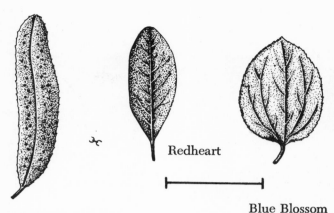

Redheart

Blue Blossom

Wartleaf Ceanothus

heart, *C. spinosus,* a very tall shrub or small tree in which the branches end in sharp spines. The dark green leaves are flexible but leathery, and usually about 1 inch long and ½ inch wide. The smooth green bark on the younger branches is distinctive, as are the large clusters of pale blue flowers that appear in late winter and early spring. Red-heart is abundant at low elevations near the coast from the vicinity of Santa Barbara to the Santa Ana Mountains. Deer Brush, *C. integerrimus,* and Palmer Ceanothus, *C. palmeri,* are not spiny and have white flowers, but both have entire leaves as does Red-heart, and are of similar size or a little larger. Both are found on mountain slopes in pine forest above 4,000 feet. Deer Brush ranges from the Santa Ana and San Jacinto mountains northward; Palmer Ceanothus, from those two ranges southward. Deer Brush commonly has leaves with 3 main veins, and the local races in the two ranges where it overlaps with Palmer Ceanothus always do. Since in Palmer Ceanothus the leaves have a single main vein, this, plus the geographical distribution, serves to separate the two species.

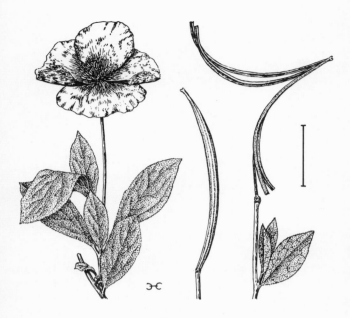

Bush Poppy

Bush Poppy *(Dendromecon rigida)*

Bush Poppy is a smooth gray-green evergreen shrub
3 to 6 feet tall, with leaves mostly about 2 to 3 inches
and one-fourth as wide with entire margins. The
yellow flowers are solitary and about 2 inches across,
and have 4 (more rarely 5) petals and numerous
stamens. The slender capsules are 2 to 4 inches long
and split upward from the bottom. Bush Poppy blooms
in spring and occurs throughout the coastal ranges of
southern California.

Toyon, California Christmas Berry *(Heteromeles
arbutifolia)*

Toyon, with its attractive large clusters of bright red
berries and dark green leaves, is commonly cultivated.
The leathery leaves are 2 to 4 inches long and about

[112]

one-third as wide, evenly and sharply toothed around the margins, and paler green beneath. The white flowers appear in summer in large terminal clusters, and have 5 petals and 10 stamens in pairs opposite the calyx-lobes. The berries are about ¼ inch long and are exactly comparable in structure to tiny apples. They appear about Christmastime and it would seem inviting to pick them, but they are rigidly protected by State law.

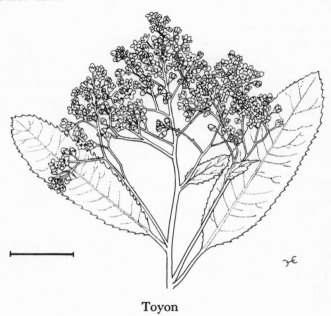

Toyon

Cliff Spray (*Holodiscus discolor*)

A low shrub with soft, hairy, toothed leaves mostly about 1½ inches long and three-fourths as wide, the tiny flowers numerous and crowded into large terminal bunches. Each flower has 5 petals and 20 stamens. Cliff Spray is found in shaded places near the coast from Orange County north, and again, in a small-

[113]

leaved form, above 6,000 feet on rocky slopes in the mountains from the San Jacintos north. It blooms in early summer.

Gray Krameria *(Krameria grayi)*

Gray Krameria is a thorny desert shrub less than 2 feet tall, with open irregular red-purple flowers about ½ inch across and pods about ¼ inch long covered with barbed prickles. The gray leaves are narrow and up to ½ inch long. Gray Krameria flowers in spring and occurs on stony or sandy slopes in the Colorado and southern Mojave deserts. Its relative, Littleleaf Krameria, *Krameria parvifolia,* which has the two upper petals joined at the base (they are entirely free in *K. grayi*), occupies the same area. If a plant has irregular flowers with free petals and simple alternate leaves and does not seem to be krameria, it should be compared with the members of the Pea Family (p. 33).

Bush Peppergrass *(Lepidium fremontii)*

A rounded desert shrub rarely more than 2 feet tall, with slender nonhairy leaves up to about 2 inches long, the lower ones divided into linear segments. The white flowers are about ⅛ inch long and gathered into terminal bunches, and have 6 stamens, the 4 outer ones longer than the 2 inner. Bush Peppergrass flowers in late winter and spring and occurs on the northern Colorado and Mojave deserts.

Bush Mallow *(Malacothamnus fasciculatus)*

Erect shrubs mostly 2 to 8 feet tall, characteristic of disturbed areas on brushy slopes and often coming up in great numbers on burned areas. The leaves are shallowly toothed and also more or less broadly and deeply lobed, somewhat maple-like, and densely covered with short hairs. The open flowers are lavender-pink to whitish and about 1 to 1½ inches across; the numerous stamens are fused into a central tube as is

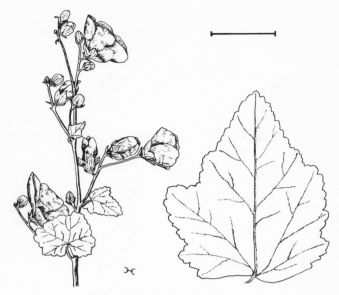

Bush Mallow

typical in members of the Mallow Family. The flowers
are either clustered into dense groups scattered along
the stem or in a more or less open arrangement. Bush
Mallow is common throughout the ranges of the
western part of our area, and reaches the margins of
the desert and moderate elevations in the mountains.

Sandpaper Plant *(Petalonyx thurberi)*

Sandpaper Plant is a spreading shrub of sandy areas,
rarely much over 2 feet in height, with rough-hairy
branches and leaves. The grayish leaves are narrowly
triangular, usually about ½ inch long, and unstalked.
The white flowers are about ⅛ inch long, and have
the unique characteristic that the stamens are exserted
between the petals so as to appear outside them. Sandpaper Plant flowers in early summer and occurs
throughout the Mojave and Colorado deserts.

[115]

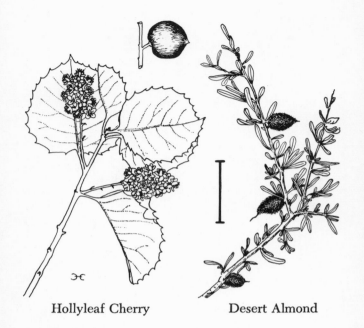

Hollyleaf Cherry   Desert Almond

Cherry, Plum, Almond *(Prunus)*

Included in this genus are a number of very diverse
southern California shrubs. One of the commonest is
the Hollyleaf Cherry, or Islay, *Prunus ilicifolia*, which
is, unlike any of the others, evergreen, with thick,
spine-toothed, holly-like leaves. These rounded leathery leaves are mostly about 1½ inches long, dark green
above and paler below. The white flowers, each about
¼ inch across, are borne in clusters 1 to 2 inches long.
The mature cherry is dark red with a very thin pulp
that is sweet when mature. Hollyleaf Cherry flowers
in spring and occurs throughout southern California
at moderate elevations away from the desert.

Two other species of *Prunus* are distinctly spiny.
Both are usually found on the desert. The Desert Almond, *P. fasciculata*, has narrow leaves less than ½

inch long clustered on short lateral branches. The flowers are white and less than ¼ inch across, and the fruit is about ⅜ inch long and densely hairy. Desert Almond occurs on the mountain slopes of both deserts, and also in northern coastal Santa Barbara County. Another spiny shrub is Desert Apricot, *P. fremontii*, which has rounded leaves about ½ to ¾ inch across and solitary white flowers about ½ inch across. The fruit is rounded and up to ½ inch long. Desert Apricot occurs along the western margin of the Colorado Desert.

The remaining two species are not spiny; both bloom in spring and occur in fairly damp places above 4,000 feet in the mountains. Bitter Cherry, *P. emarginata*, is an erect shrub 4 to 12 feet tall or even a small tree, with leaves ¾ inch to 2 inches long and half as wide, the margins minutely toothed. The white flowers are about ½ inch across and occur in small clusters of 3 to 10. Western Choke-cherry, *P. virginiana*, is similar in habit, but has somewhat larger leaves, and its white flowers are borne in elongate clusters 2 to 5 inches long which terminate the branches instead of being lateral as in Bitter Cherry.

Bitterbrush *(Purshia tridentata)*

Erect desert shrub, often 4 to 6 feet tall, with densely clustered, deeply lobed leaves ¼ to ½ inch long. The pale yellow flowers are about ½ inch across and are scattered among the leaves. Bitterbrush flowers in spring and occurs from the western margin of the Colorado Desert to the Mojave Desert and Mount Pinos.

California Coffee Berry *(Rhamnus californica)*

Erect evergreen shrub with oblong leaves 1 to 3 inches long and ½ to 1 inch wide, dark green above and pale green or grayish with short hairs below. The leaves vary from entire to finely toothed all around. The tiny

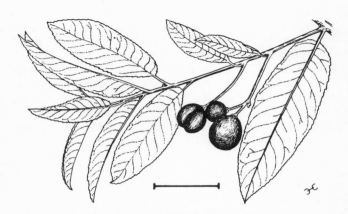

California Coffee Berry

greenish flowers are borne in small lateral clusters and are followed by subglobose berries ¼ inch or more thick, which are at first green, later red, and finally lustrous black, with 2 or 3 large seeds enclosed. California Coffee Berry has a number of distinctive local races, and one or more of these exist throughout the lower and middle elevations of southern California, extending to the desert mountains.

Lemonade Berry *(Rhus integrifolia)*

A stout evergreen shrub with rounded, leathery dark green leaves mostly 1 to 2 inches long and ¾ inch to 1½ inches wide, paler green beneath. The leaves are entire or irregularly toothed around the margins. The white or pinkish flowers have 5 petals and 5 stamens and are borne in compact terminal branched clusters. The flattened reddish fruits are densely covered with sticky bumps or short hairs. Lemonade Berry occurs along the coast and coastal mountains of southern California, blooming in late winter. The individual shrubs are spreading or mostly up to a few feet tall, sometimes taller.

[118]

Sugarbush *(Rhus ovata)*

Sugarbush is similar in stature, habit, and leaf size to Lemonade Berry, and has comparable fruits and flowers, the latter appearing in early spring. The leaves, however, are often trough-like above, always entire, and more pointed at the apex, and their stalks are over ⅜ inch long, instead of shorter, as in Lemonade Berry. Sugarbush inhabits approximately the same range as Lemonade Berry, but is usually at higher elevations and extends east to the margins of the desert.

Laurel Sumac *(Rhus laurina)*

A tall leafy evergreen shrub with leaves 2 to 4 inches long and ¾ inch to 2 inches wide, entire, long-stalked, and usually pointed at the apex. While still leathery,

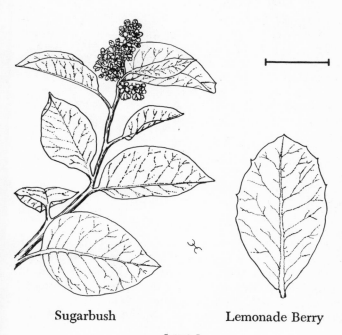

Sugarbush   Lemonade Berry

[119]

its leaves are thinner than those of the two preceding species. The flowers are smaller and appear in early summer in dense terminal clusters. The fruits are white, smooth, and subglobose, and about ⅛ inch in diameter. Laurel Sumac occurs throughout the lower mountain slopes of the western part of our area.

Currant, Gooseberry *(Ribes)*

The fairly numerous species of currant and gooseberry in southern California can be divided rather easily into two groups: the currants, which lack spines and prickles; and the gooseberries, which have spines and often prickles too. All are deciduous except *Ribes speciosum,* a gooseberry.

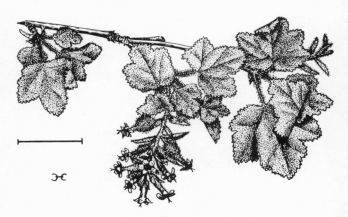

Chaparral Flowering Currant

CURRANTS

Of the four common kinds of currant in our area, Western Golden Currant, *Ribes aureum,* is the only one that has yellow flowers. It is an erect shrub with rather firm, deeply 3-lobed, grayish leaves and small lateral clusters of yellow flowers about ½ inch long. As in all species of *Ribes,* the calyx is the conspicuous part of the flower and the petals are only about ⅛ inch long. The smooth globose berries are about ¼ inch thick and range in color from yellow to black. Western Golden Currant blooms in late winter and spring, and occurs from the San Bernardino and Santa Monica mountains north in the Coast Ranges away from the immediate coast. The other three southern California currants have pink or purplish to white flowers. Squaw Currant, *R. cereum,* has rounded tough leaves less than 1 inch across which are often rather shallowly lobed. The pinkish to white flowers are about ¼ inch long and in short lateral clusters. Squaw Currant blooms in early summer and occurs on dry slopes above 6,000 feet in the mountains. Chaparral Flowering Currant, *R. malvaceum,* likewise has purplish or white flowers, but they are about ½ inch long or a little less and are borne in somewhat drooping clusters that appear in winter with the first leaves. The lobed, maple-like leaves of this species are dark green and roughened above, paler below, and mostly about 2 inches across (smaller in some of the white-flowered forms). The smooth or hairy blue berries are about ¼ inch in diameter. *Ribes malvaceum* occurs on dry slopes below 5,000 feet throughout the coastal ranges. The fourth fairly common currant, Mountain Pink Currant, *R. nevadense,* grows in moist places, as along streams in our higher mountains above 5,000 feet. It is in general similar to the preceding, but easily distinguished by its habitat.

GOOSEBERRIES

The gooseberries are a distinctive group, with their spiny branches and deeply lobed leaves, which are rounded and mostly ½ to 1 inch across. Our most distinctive species is Fuchsia Flowering Gooseberry, *Ribes speciosum,* which has bright red flowers hanging under the branches and often covering them. It is not only our only evergreen member of the genus, it is the only one with 4 sepals and petals instead of 5, and also our only species with bright red flowers. This species should easily be recognized by reference to plate 7. It occurs throughout the mountains of the western part of our area. Most of the remaining species of goosebery have elongate calyx-tubes, but one, Mountain Gooseberry, *R. montigenum,* has saucer-shaped green flowers, becoming reddish with age, about ¼ inch across. It is a rather rare low spreading shrub found on rocky slopes, usually above 9,000 feet in the mountains.

The remaining two gooseberries have tubular purplish or greenish flowers about ⅜ inch long with white petals. They are both low shrubs, often about 3 feet tall. Sierra Gooseberry, *R. roezlii,* is a very spiny shrub that is fairly frequent in the mountains above 4,500 feet, where it often grows in open pine forest. Its berries are about ½ inch in diameter and densely covered with stout spines. Hillside Gooseberry, *R. californicum,* is similar in general but occurs mostly below 2,000 feet, from the south face of the San Gabriel Mountains and the Santa Monicas north in the coastal ranges.

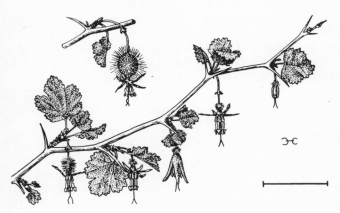

Hillside Gooseberry

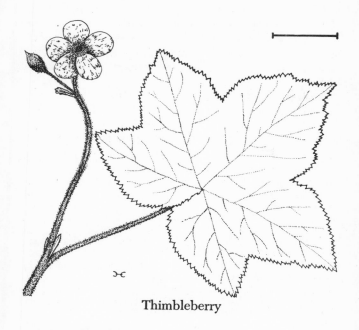

Thimbleberry

Thimbleberry *(Rubus parviflorus)*

The tasty pinkish fruits of Thimbleberry are well known to campers in our mountain regions. These fruits are about ½ to ¾ inch across, and easily separate in a thimble-shaped berry from their stalk. The maple-like leaves are mostly 3 to 6 inches across and usually densely covered with soft hairs. The white flowers, which appear in late spring, are 1 to 2 inches across and have 5 petals and numerous stamens. Thimbleberry is found in shaded ravines and along streams, mostly above 5,000 feet in the mountains of the western part of our area.

Turpentine Broom *(Thamnosma montana)*

This low, rigid shrub of the Colorado and Mojave deserts, with its aromatic yellowish-green branches, is totally unlike any other southern California shrub. The entire leaves are early deciduous and less than ½ inch long; the 4-petaled purple flowers are elongate and about ¼ to ½ inch long. They appear in spring. The dry fruit is 2-lobed and hence broadly heart-shaped.

SUGGESTED REFERENCES

McMinn, Howard E. *An Illustrated Manual of California Shrubs*. Berkeley: University of California Press, 1939.

Munz, Philip A., in collaboration with David D. Keck. *A California Flora*. Berkeley and Los Angeles: University of California Press, 1959.

Abrams, L. R. *Illustrated Flora of the Pacific States*. 4 vols. Stanford: Stanford University Press, 1823-1960.

Jaeger, E. C. *Desert Wild Flowers*. Stanford: Stanford University Press, 1940.

Smith, Clifton F. *A Flora of Santa Barbara*. Santa Barbara: Santa Barbara Botanic Garden, 1952.

Higgins, Ethel B. *Annotated Distributional List of the Ferns and Flowering Plants of San Diego County, California*. San Diego Society of Natural History, 1949.

CHECK LIST OF
SOUTHERN CALIFORNIA NATIVE SHRUBS

CYPRESS FAMILY (CUPRESSACEAE)
Juniperus californica, California Juniper, p. 17

EPHEDRA FAMILY (EPHEDRACEAE)
Ephedra californica, p. 16
Ephedra nevadensis, p. 16
Ephedra trifurca, p. 17
Ephedra viridis, p. 16

LILY FAMILY (LILIACEAE)
Nolina bigelovii, Bigelow Nolina, p. 18
Nolina interrata, p. 19
Nolina parryi, Parry Nolina, p. 19
Nolina wolfii, Wolf Nolina, p. 19
Yucca schidigera, Mojave Yucca, p. 19

WILLOW FAMILY (SALICACEAE)
Salix hindsiana, Sandbar Willow, p. 104
Salix lasiolepis, Arroyo Willow, p. 104
Salix lutea, Yellow Willow, p. 105
Salix scouleriana, Scouler Willow, p. 104

OAK FAMILY (FAGACEAE)
Castanopsis sempervirens, Bush Chinquapin, p. 96, pl. 6
Quercus dumosa, California Scrub Oak, p. 101
Quercus durata, Leather Oak, p. 101
Quercus kelloggii, California Black Oak, p. 101
Quercus palmeri, Palmer Oak, p. 101
Quercus wislizenii, Interior Live Oak, p. 102, pl. 7

WALNUT FAMILY (JUGLANDACEAE))
Juglans californica, Southern California Black Walnut, p. 28

MISTLETOE FAMILY (LORANTHACEAE)
Phoradendron bolleanum, Cypress Mistletoe, p. 24
Phoradendron californicum, Mesquite Mistletoe, p. 24
Phoradendron juniperinum, Juniper Mistletoe, p. 24
Phoradendron tomentosum, Bigleaf Mistletoe, p. 24

SWEET GALE FAMILY (MYRICACEAE)
Myrica californica, California Wax Myrtle, p. 99
Phoradendron villosum, Oak Mistletoe, p. 24

BUCKWHEAT FAMILY (POLYGONACEAE)
Eriogonum cinereum, Ashleaf Buckwheat, p. 97
Eriogonum fasciculatum, California Buckwheat, p. 98, pl. 6
Eriogonum parvifolium, Seacliff Buckwheat, p. 98

SALTBUSH FAMILY (CHENOPODIACEAE)
Allenrolfea occidentalis, Bush Pickleweed, p. 95
Atriplex canescens, Wingscale, p. 95
Atriplex confertiflora, Shadscale, p. 96
Atriplex hymenelytra, Desert Holly, p. 95, pl. 6
Atriplex lentiformis, Lenscale, p. 96
Atriplex parryi, Parry Saltbush, p. 95
Atriplex polycarpa, Allscale, p. 95
Atriplex spinifera, Spinescale, p. 96
Eurotia lanata, Winter Fat, p. 99
Grayia spinosa, Grayia, p. 99
Suaeda californica, California Sea Blight, p. 105

BUTTERCUP FAMILY (RANUNCULACEAE)
Clematis lasiantha, Pipe-stem Clematis, p. 19
Clematis ligusticifolia, Western Clematis, p. 21
Clematis pauciflora, Southern California Clematis, p. 21

BARBERRY FAMILY (BERBERIDACEAE)
Berberis dictyota, California Barberry, p. 26
Berberis nevinii, Nevin Barberry, p. 26
Berberis pinnata, Shinyleaf Barberry, p. 25, pl. 1

POPPY FAMILY (PAPAVERACEAE)
Dendromecon rigida, Bush Poppy, p. 112

CAPER FAMILY (CAPARIDACEAE)
Isomeris arborea, Bladderpod, p. 27, pl. 1

MUSTARD FAMILY (CRUCIFERAE)
Lepidium fremontii, Bush Peppergrass, p. 114

SAXIFRAGE FAMILY (SAXIFRAGACEAE)
Ribes aureum, Western Golden Currant, p. 120
Ribes californicum, Hillside Gooseberry, p. 122
Ribes cereum, Squaw Currant, p. 121
Ribes malvaceum, Chaparral Flowering Currant, p. 121
Ribes montigenum, Mountain Gooseberry, p. 122
Ribes nevadense, Mountain Pink Currant, p. 121
Ribes roezlii, Sierra Gooseberry, p. 122
Ribes speciosum, Fuchsia Flowering Gooseberry, p. 121

ROSE FAMILY (ROSACEAE)

Adenostoma fasciculatum, Chamise, p. 106, pl. 7
Adenostoma sparsifolium, Red Shanks, p. 106, pl. 7
Amelanchier alnifolia, Service Berry, p. 107
Cerocarpus ledifolius, Desert Mountain Mahogany, p. 97
Cercocarpus montanus, Western Mountain Mahogany, p. 97
Chamaebatia australis, Southern Mountain Misery, p. 27
Coleogyne ramosissima, Blackbrush, p. 91
Heteromeles arbutifolia, Toyon, California Christmas Berry, p. 112, pl. 8
Holodiscus discolor, Cliff Spray, p. 113
Prunus emarginata, Bitter Cherry, p. 117
Prunus fasciculata, Desert Almond, p. 117
Prunus fremontii, Desert Apricot, p. 117
Prunus ilicifolia, Hollyleaf Cherry, Islay, p. 116
Prunus virginiana, Western Choke-cherry, p. 117
Purshia tridentata, Bitterbush, p. 117, pl. 8
Rosa californica, California Rose, p. 31
Rosa gymnocarpa, Wood Rose, p. 31
Rosa woodsii, Mountain Rose, p. 31
Rubus leucodermis, Western Raspberry, p. 32
Rubus parviflorus, Thimbleberry, p. 123
Rubus procerus, Himalaya Berry, p. 32
Rubus ursinus, Wild Blackberry, p. 31

PEA FAMILY (LEGUMINOSAE)

Acacia greggii, Cat's Claw, p. 35
Amorpha californica, California False Indigo, p. 36
Calliandra eriophylla, Fairy Duster, p. 37
Cassia armata, Armed Senna, p. 37
Cercis occidentalis, Western Redbud, p. 37
Dalea arborescens, Mojave Dalea, p. 38
Dalea emoryi, Emory Dalea, p. 38
Dalea fremontii, Fremont Dalea, p. 39
Dalea polyadenia, Nevada Dalea, p. 39
Dalea schottii, Mesa Dalea, p. 39, pl. 2
Dalea spinosa, Smoke Tree, p. 39
Hoffmanseggia microphylla, Rushpea, p. 40
Lotus rigidus, Desert Rock-pea, p. 40
Lotus scoparius, Deerweed, p. 40
Lupinus albifrons, Silver Lupine, p. 41
Lupinus arboreus, Tree Lupine, p. 41
Lupinus chamissonis, Dune Lupine, p. 41
Lupinus longifolius, Pauma Lupine, p. 41
Pickeringia montana, Chaparral Pea, p. 41
Prosopis glandulosa, Mesquite, p. 42
Prosopis pubescens, Screwbean, p. 43

MILKWORT FAMILY (POLYGALACEAE)
Krameria grayi, Gray Krameria, p. 114
Krameria parvifolia, Littleleaf Krameria, p. 114

SPURGE FAMILY (EUPHORBIACEAE)
Ricinus communis, Castor Bean, p. 102

CALTROPS FAMILY (ZYGOPHYLLACEAE)
Larrea tridentata, Creosote Bush, p. 28, pl. 2

RUE FAMILY (RUTACEAE)
Cneoridium dumosum, Bushrue, p. 90
Thamnosma montana, Turpentine Broom, p. 123

BOX FAMILY (BUXACEAE)
Simmondsia chinensis, Jojoba, Goatnut, p. 94

SUMAC FAMILY (ANACARDIACEAE)
Rhus diversiloba, Poison Oak, pp. 22, 30, pl. 1
Rhus integrifolia, Lemonade Berry, p. 118
Rhus laurina, Laurel Sumac, p. 119, pl. 8
Rhus ovata, Sugarbush, p. 119
Rhus trilobata, Squaw Bush, p. 30, pl. 2

STAFF-TREE FAMILY (CELASTRACEAE)
Euonomys occidentalis, Western Burning Bush, p. 92

MAPLE FAMILY (ACERACEAE)
Acer glabrum; Dwarf Maple, Mountain Maple, p. 90

BUCKTHORN FAMILY (RHAMNACEAE)
Adolphia californica, California Adolphia, p. 90
Ceanothus cordulatus, Snow Bush, p. 108
Ceanothus crassifolius, Hoaryleaf Ceanothus, p. 107
Ceanothus cuneatus, Buckbrush, p. 108
Ceanothus cyaneus, San Diego Ceanothus, p. 110
Ceanothus foliosus, Wavyleaf Ceanothus, p. 110
Ceanothus greggii, p. 108.
Ceanothus integerrimus, Deer Brush, p. 110
Ceanothus leucodermis, Chaparral Whitehorn, p. 108
Ceanothus megacarpus, Bigpod Ceanothus, p. 108
Ceanothus oliganthus, Hairy Ceanothus, p. 110
Ceanothus palmeri, Palmer Ceanothus, p. 111
Ceanothus papillosus, Wartleaf Ceanothus, p. 110, pl. 7
Ceanothus spinosus, Red-heart, p. 111

Ceanothus thyrsiflorus, Blue Blossom, p. 110
Ceanothus verrucosus, Wartystem Ceanothus, p. 108
Rhamnus californica, California Coffee Berry, p. 117
Rhamnus crocea, Redberry, p. 102

GRAPE FAMILY (VITACEAE)
Vitis girdiana, Wild Grape, p. 23

MALLOW FAMILY (MALVACEAE)
Malacothamnus fasciculatus, Bush Mallow, p. 114

BOMBAX FAMILY (BOMBACEACEAE)
Fremontodendron californicum, Fremontia, p. 99

TAMARIX FAMILY (TAMARICACEAE)
Tamarix pentandra, p. 17
Tamarix tetrandra, p. 17

OCOTILLO FAMILY (FOUQUIERIACEAE)
Fouquieria splendens, Ocotillo, p. 83, pl. 5

LOASA FAMILY (LOASACEAE)
Petalonyx thurberi, Sandpaper Plant, p. 115

OLEASTER FAMILY (ELEAGNACEAE)
Shepherdia argentea, Silver Buffaloberry, p. 94

GARRYA FAMILY (GARRYACEAE)
Garrya elliptica, p. 93
Garrya flavescens, Pale Silktassel, p. 92
Garrya veatchii, Veatch Silktassel, p. 93

DOGWOOD FAMILY (CORNACEAE)
Cornus glabrata, Smooth Dogwood, p. 92
Cornus occidentalis, Creek Dogwood, p. 92

HEATHER FAMILY (ERICACEAE)
Arctostaphylos glandulosa, Eastwood Manzanita, p. 78
Arctostaphylos glauca, Bigberry Manzanita, p. 78
Arctostaphylos parryana, Parry Manzanita, p. 78
Arctostaphylos pringlei, Pinkbract Manzanita, p. 78
Arctostaphylos pungens, Mexican Manzanita, p. 78
Arctostaphylos viridissima, Lompoc Manzanita, p. 78
Comarostaphylis diversifolia, Summer Holly, p. 80, cover
Gaultheria shallon, Salal, p. 83
Rhododendron occidentale, Western Azalea, p. 87, pl. 5
Vaccinium ovatum, California Huckleberry, p. 88
Xylococcus bicolor, Mission Manzanita, p. 88

[129]

STORAX FAMILY (STYRACACEAE)
Styrax officinalis, Snowdrop Bush, p. 88

OLIVE FAMILY (OLEACEAE)
Fraxinus dipetala, Chaparral Flowering Ash, p. 27, pl. 1
Menodora spinescens, Spiny Mendora, p. 86

PHLOX FAMILY (POLEMONTACEAE)
Leptodactylon californicum, Prickly Phlox, p. 83, pl. 5

PHACELIA FAMILY (HYDROPHYLLACEAE)
Eriodictyon capitatum, Lompoc Yerba Santa, p. 82
Eriodictyon crassifolium, Thickleaf Yerba Santa, p. 81
Eriodictyon denudatum, Santa Barbara Yerba Santa, p. 81
Eriodictyon lanatum, San Diego Yerba Santa, p. 81
Eriodictyon traskiae, Trask Yerba Santa, p. 81
Eriodictyon trichocalyx, Smooth Yerba Santa, p. 81

MINT FAMILY (LABIATAE)
Hyptis emoryi, Desert Lavender, p. 61
Lepechinia, Pitcher Sage, p. 62
Salazaria mexicana, Bladder Sage, p. 69, pl. 3
Salvia apiana, White Sage, p. 71
Salvia brandegei, p. 74
Salvia clevelandii, Cleveland Sage, p. 73, pl. 4
Salvia dorrii, Desert Sage, p. 70, pl. 4
Salvia eremostachya, Sand Sage, p. 71
Salvia greatae, Greata Sage, p. 70
Salvia leucophylla, Purple Sage, p. 73
Salvia mellifera, Black Sage, p. 73
Salvia mohavensis, Mojave Sage, p. 72
Salvia munzii, p. 74
Salvia pachyphylla, Mountain Desert Sage, p. 70
Salvia vaseyi, Bristle Sage, p. 72
Trichostema lanatum, Woolly Blue-curls, p. 74, pl. 4
Trichostema parishii, Parish Blue-curls, p. 75

NIGHTSHADE FAMILY (SOLANACEAE)
Lycium andersonii, Anderson Desert Thorn, p. 86
Lycium brevipes, Desert Thorn, p. 85
Lycium californicum, Coast Desert Thorn, p. 85
Lycium cooperi, Cooper Desert Thorn, p. 84
Lycium fremontii, Fremont Desert Thorn, p. 86
Lycium pallidum, Rabbit Thorn, p. 84
Lycium torreyi, Torrey Desert Thorn, p. 86
Nicotiana glauca, Tree Tobacco, p. 87, pl. 5
Solanum xanti, Chaparral Nightshade, p. 87, pl. 6

FIGWORT FAMILY (SCROPHULARIACEAE)
Mimulus aridus, p. 68
Mimulus aurantiacus, Northern Monkeyflower, p. 67
Mimulus calycinus, p. 66
Mimulus longiflorus, Southern Monkeyflower, p. 66
Mimulus puniceus, Red Monkeyflower, p. 65
Penstemon antirrhinoides, Yellow Penstemon, p. 68
Penstemon breviflorus, p. 69
Penstemon cordifolius, Climbing Penstemon, p. 68
Penstemon rothrockii, p. 69
Penstemon ternatus, Whorl-leaf Penstemon, p. 68

ACANTHUS FAMILY (ACANTHACEAE)
Beloperone californica, California Beloperone, p. 61, pl. 3

BIGNONIA FAMILY (BIGNONIACEAE)
Chilopsis linearis, Desert Willow, p. 79, pl. 4

MADDER FAMILY (RUBIACEAE)
Galium angustifolium, Chaparral Bedstraw, p. 92
Galium stellatum, p. 92

HONEYSUCKLE FAMILY (CAPRIFOLIACEAE)
Lonicera hispidula, California Honeysuckle, p. 63
Lonicera interrupta, Chaparral Honeysuckle, p. 63
Lonicera involucrata, Twinberry, p. 62
Lonicera subspicata, Southern Honeysuckle, p. 63
Sambucus caerulea, Blue Elderberry, p. 33
Sambucus microbotrys, p. 33
Symphorocarpos albus, Common Snowberry, p. 74
Symphoricarpos vaccinioides, Mountain Snowberry, p. 74

SUNFLOWER FAMILY (COMPOSITAE)
Acamptopappus sphaerocephalus, Goldenhead, p. 49
Artemisia arbuscula, p. 54
Artemisia californica, Coast Sagebrush, p. 53
Artemisia palmeri, Tall Sagebrush, p. 53
Artemisia rothrockii, Rothrock Sagebrush, p. 54
Artemisia spinescens, Bud Sagebrush, p. 53
Artemisia tridentata, Sagebrush, p. 53
Baccharis douglasii, p. 57
Baccharis emoryi, p. 58
Baccharis gultinosa, Seep Willow, p. 57
Baccharis pilularis, Coyote Brush, p. 57
Baccharis plummerae, p. 56
Baccharis sarothroides, Broom Baccharis, p. 56
Baccharis sergiloides, Squaw Waterweed, p. 57